GODS, PHILOSOPHERS, and SCIENTISTS

RELIGION AND SCIENCE IN THE WEST

SCOTT E. HENDRIX

OXFORD SOUTHERN

an imprint of Sunbury Press, Inc.
Mechanicsburg, PA USA

OXFORD SOUTHERN

an imprint of Sunbury Press, Inc.
Mechanicsburg, PA USA

For information about special discounts for bulk purchases, please contact Sunbury Press Orders Dept. at (855) 338-8359 or orders@sunburypress.com.

To request one of our authors for speaking engagements or book signings, please contact Sunbury Press Publicity Dept. at publicity@sunburypress.com.

ISBN: 978-1-62006-335-4 (Trade paperback)

Library of Congress Control Number: 2019948920

FIRST OXFORD SOUTHERN EDITION: January 2020

Product of the United States of America
0 1 1 2 3 5 8 13 21 34 55

Set in Bookman Old Style
Designed by Crystal Devine
Cover by Terry Kennedy
Edited by Erika Hodges

Continue the Enlightenment!

Dedication

As with every author, I owe a great many people thanks for assistance, inspiration, and support over the years in which I thought about, developed, and wrote this book, beginning with my mother, Dorothy Hendrix. I grew up in poverty in rural Alabama, but she always made sure we had a home filled with books. And though her education had stopped in the 5th grade when her parents took her out of school to work on the family farm, she read to me constantly when I was a child, encouraged my intellectual interests, and set me on the path to become an academic, even if I followed a very winding road to get there.

Speaking of academics, there have been a great number who have had an impact on me. My dissertation advisor at the University of Tennessee, Tom Burman, always pushed me to be the best I could be, while generously supporting me both in the classroom and out—even to the point of working with me during his free time on medieval Latin and paleography. At Carroll University, Kevin Guilfoy was kind enough to give me the opportunity to teach "The History and Philosophy of Science," and we've had numerous fruitful discussions about the philosophy of science as well as many other things over the years. This book grew out of my experiences teaching this class, and many of the ideas are things he and I have discussed over beer. Some years ago I had the pleasure of meeting Uchenna Okeja, now at the University of Rhodes in South Africa, at a conference in Salzburg, leading to multiple collaborations on writing projects. Uchenna is a philosopher whose work on post-secularism has influenced my thought greatly. Finally, I should mention that I owe a debt of gratitude to Carroll University for the support I've received over the years in terms of research funding and that most important resource, time. I wrote most of this book while on sabbatical.

Speaking of support, my wife, Kelly Hendrix, has been steadfast in her encouragement over the years. It wouldn't surprise me if she and our dogs know the material in this book as well as I do, given the number of times I've talked it over with her as we've walked our pups. I couldn't ask for a better partner than Kelly, and I appreciate her patience with me as I researched and wrote this book, as well as all the other things I've written. I know I get a bit stressed out and cranky sometimes when I'm immersed in a project.

Finally, I would like to thank the good people at Sunbury Press for believing in this project and bringing it to fruition. I told Lawrence Knorr, the founder and CEO, via email that I was drawn to the press because of the motto, "Continue the Enlightenment." Based on my experiences working with great people like my editor, Erika Hodges, I would say the Enlightenment is in good hands with Sunbury.

Introduction

For some years, I've taught a class with the rather unimaginative title, "The History and Philosophy of Science." This is a seminar-style course, meaning that it's a rare day when I lecture. Instead, I generally assign readings to the students and ask questions that I hope are probing and meant to provoke critical thought. Where I teach, most students haven't encountered this style of teaching before, so I ease them into it on the first day of class by asking questions meant to get them thinking about what they do and don't know about the subject. The trick is to guide the discussion down certain alleyways of thought without making it obvious that I'm actually doing any guiding. Instead, I want the students to feel that it's a fully free-form discussion where they get to share and develop their thoughts—and to some degree it is. But there are certain questions that I want to raise, such as "what is science?" and "what is the relationship between science and religion?" Questions like that aren't nearly as straightforward as they seem. And they're connected to two of my goals during the semester, which is to help the students understand the complex nature of science, and the equally complex relationship between science and religion. From the first time I taught this class, I noticed that students were frequently baffled to learn that throughout history religious beliefs inspired and contributed to scientific understanding in the West at least as often as they hindered scientific development. That may surprise many readers, and that surprise is why I decided to write this book. In the following chapters, I will not only explain why and how religious thought in the West has normally had a symbiotically productive relationship with scientific approaches to understanding the world, but I will also seek to clarify why the mistaken idea that science and religion are inherently antagonistic developed.

As we'll see in the chapters to come, my argument isn't one based on ideology or religious sentiment, both of which I avoid as much as possible. Instead, it's based on a concrete evaluation of the historical record, even if it's not one that most people will intuitively accept. After all, most Americans agree with my students that religion and science don't play well together—a 2014 Pew Research Center survey reports that 59% of American adults believe religion and science are often in conflict, while only 38% of respondents state that the two are mostly compatible. Interestingly, it seems that people who are the least religious are also the most likely to believe that science and religion are often in conflict, with 76% of adults self-identifying as not being affiliated with any religious tradition saying that religious and scientific views are often in conflict. On the other hand, 65% of black Protestants agree, running a close second to the religiously unaffiliated. But this group is something of an outlier, for white and Hispanic evangelicals and Catholics are split roughly 50/50 on the question of whether religion and science are in conflict or are compatible, which is also true of white evangelical Protestants. Perhaps most intriguingly, only 30% of American adults say their own religious beliefs conflict with science, while 68% state that their personal religious beliefs and science don't conflict.

The last paragraph threw many numbers at you, something I won't do regularly. So what do all those numbers mean? To put it simply, most Americans believe that religion and science are incompatible, an attitude that is especially prevalent among those who are religiously unaffiliated. In 2014 the Pew Research Center also conducted a "Religious Landscape Study," and found that a large percentage of those who are termed "nones" (because on survey forms under "religious preference" they mark "none") have negative attitudes about religion in general or organized religion more specifically. Almost half state they have no religious beliefs, a number that includes many who state they moved away from religion because of "science" or "scientific beliefs" and see religion as incompatible with science. A full 20% of "nones" state they dislike organized religion for reasons ranging from stories of clerical sex abuse or that religion has become too much like a business, to a feeling that religious beliefs promote small or closemindedness. Moving beyond the "nones," though, to those who state they do possess a religious affiliation, most of those people who see religion and science as incompatible don't think their personal religious beliefs conflict with science. For most of these religious people, it is some ill-defined group of *other* religious people who hold religious views that conflict with science. In other words, to simplify the discussion, there are an awful lot of American adults who believe religion and science conflict *for other people*, but not for them. Why? Because of a general narrative that is widely accepted, which states that religion and science have always been in conflict, and *must* conflict by the very nature of these different approaches to truth. It's a bit odd that so many people hold that belief, when it is personally untrue for so many respondents to the Pew surveys, which raises the question of why this belief is so prevalent.

One reason is because of the large number of prominent public intellectuals who feel the conflict is a very keen one. The most well-known writers in this group are the so-called "Four Horsemen of the Non-Apocalypse," the evolutionary biologist Richard Dawkins (born in 1941), the journalist and author Christopher Hitchens (1941–2011), the neuroscientist Sam Harris (born in 1967) and the philosopher Daniel Dennett (born in 1942). While these men, and others such as the theoretical physicist Lawrence Krauss (born in 1954), differ in their approach, they all share a common antipathy toward religion—although we'll need to consider what they mean by "religion" in a bit— and a belief that science is actively under attack by those who hold religious beliefs. For writers such as these, science is not merely a superior path toward knowledge, it is the only legitimate pathway to actually knowing anything. Sure, one or two might give a nod toward the usefulness of philosophy, but none of these writers would acknowledge that theology has anything useful to offer. In fact, in his 2004 work, *The End of Faith,* Sam Harris called theology "ignorance with wings." Even that statement is mild compared to other views his colleagues sometime express. For example, during a 1992 lecture at the Edinburgh International Science Festival, Richard Dawkins stated that "faith is the great cop-out, the great excuse to evade the need to think and evaluate evidence." He also reportedly said, "I am against religion because it teaches us to be satisfied with not understanding the world."

It's important to note that the viewpoint of such writers isn't just reflective of the current majority opinion in the United States, that religion and science must by necessity conflict. They are also drivers of that viewpoint. Among the so-called New Atheists—which includes, but isn't limited to, the so-called "Four Horsemen"— Christopher Hitchens alone has sold millions of books worldwide. We need look no further than the title of his most popular work, *God Is Not Great: How Religion Poisons Everything,* to get a flavor of the attitude he promotes. Published in 2007, this book rocketed to the number two spot on Amazon's bestseller list a week after publication (right behind *Harry Potter and the Deathly Hallows,* in fact) and in its third week of publication made it to the number three spot on the *New York Times* bestseller list. Hitchens' argument is rather simple, some might in fact say it's simplistic, and is more or less the same argument found in most of his thirty books, not to mention the daily columns he wrote. He argues that organized religion is "violent, irrational, intolerant, allied to racism, tribalism, and bigotry, invested in ignorance and hostile to free inquiry, contemptuous of women and coercive toward children." For Hitchens, science is everything and religion is worse than nothing.

While Harris, Hitchens, and Dawkins are rather extreme in their expressed dislike of (or open contempt for) religion, they are far from the only ones to express the viewpoint that religion and science conflict in a popular venue. The most popular television shows ever to deal with science and its history are the two iterations of the television series *Cosmos* produced and aired by PBS. The first—with the full title, *Cosmos: A Personal Voyage*— was narrated by and based on the book of the same name by the astrophysicist Carl Sagan (1934–1996). The recipient of two Emmys and a Peabody award. Sagan's *Cosmos* has introduced more than 500 million people in sixty countries to the history of science since it was first broadcast in 1980. In the more recent 2014 version—with the full title of *Cosmos: A Spacetime Odyssey*— Neil deGrasse Tyson (1958–present)—another astrophysicist—took up the role Sagan played earlier. If anything, this second iteration of the series was even more successful than the first, reaching more than 400 million people in over sixty countries in its first broadcast run. And what do these shows tell viewers about the historical relationship between religion and science? In both of these televised explorations of the history of science, we are told repeatedly that scientists are those who "question authority" in order to uncover empirically verifiable truths. As for religion, the third episode of the 2014 version of the series is perhaps the best example of how these two series portray religious beliefs. Titled "When Knowledge Conquered Fear," the episode focuses on the discovery and publication of Sir Isaac Newton's (1643–1727) laws of motion, which "decoupled the motions of the heavens from their ancient connections to our fears" by allowing people to understand that comets are natural phenomena rather than portents of misfortune. More broadly, the message of the episode—and the series—is that science allows people to understand that the universe is mechanistic rather than the product of some divine plan, as scientists uncovered truths about the universe that toppled the superstitions people held about nature. In this version of the history of science, science represents knowledge and religion represents fearful superstition. Who wouldn't prefer the former to the latter?

As historians of science, it turns out that both Sagan and Tyson are excellent scientists. From time to time, we'll come back to the two versions of *Cosmos* they hosted in order to address some ways they get the history—and the historical relationship between religion and science—wrong. However, that should not be seen as an indication that these television shows or these scientists do their jobs badly. In fact, it is the very success of these shows and the importance of the work Sagan and Tyson have done that will make the two versions of *Cosmos* something of a touchstone in the pages that follow. Hundreds of millions of people around the world—myself included—received their first (and in many cases only) taste of the history of science from these two men. And both of them are not only highly reputable scientists, but also brilliant communicators. However, they are also both embedded in their culture (as we all are) as well as the culture of their profession, which affects the views they express in ways that they—and their viewers—are likely unaware.

For example, a running thread in both iterations of *Cosmos*, as well as in the work of the New Atheists I discussed above, is about the relationship between religion and science. But what those terms mean, as well as important elements that need to be understood if we are going to be able to dig into that relationship, aren't considered fully enough. To understand what that means, let's start with the term "religion." While somewhat extreme in the way he expresses himself, the work of Christopher Hitchens provides important insights about how this term is commonly understood whenever anyone raises a question about the relationship between science and religion. Although writers like Hitchens are not always completely upfront about their target, it's fairly clear to even the casual reader that the Abrahamic faiths—Judaism, Christianity, and Islam—are what they mean when they discuss religion. Furthermore, Christianity receives most of their attention, likely because it's the dominant religious faith of the culture that produced these writers. Sure, Hitchens makes a few passing references to Buddhism and Hinduism—and his fellow writer Sam Harris has many negative things to say about Islam—but for the most part, when writers like Hitchens say "religion" what they really mean is Christianity. And in order to make these attacks, they typically focus on areas of contention such as evolutionary theory or cosmological ideas such as the Big Bang theory, relying heavily on personal anecdotes (at least in the case of Hitchens), the scientific case for these theories, or quasi-philosophical arguments about how we come to know things. But these aren't just attacks on Christianity—they are, in the minds of their authors, counterattacks in a war between religion and science. In their view, religion—or at least Christianity—has always opposed science, and Christians have attempted to destroy science in the past and continue their efforts in the present.

Thus, when writers who focus on the relationship between science and religion use the term "religion," what they usually mean is Christianity. Any attempt to define what religion might be is therefore lacking. That's too bad, because there are many thoughtful studies about this topic. For example, Robert Crawford gives a bit of an overview of the difficulties of providing a definition in his 2002 study, *What is Religion?* He points to the complexities of providing a singular definition, and highlights the important functional aspects of religion relating to a belief in the divine that binds a social group together while providing a moral code (among other factors). Crawford's definition is intentionally broad, for he is thinking of polytheistic, animistic, and atheist religions such as Buddhism, as well as the monotheistic religions with which Americans are most familiar. Perhaps such a philosophical consideration of what one means by the term religion isn't necessary to the work people such as Hitchens, Harris, or Dawkins are doing, but it would be useful if they expressed more of an awareness that religion and Abrahamic monotheism aren't synonymous. Carl Sagan is actually better about dealing with this issue—although not by much—in his 1980 version of *Cosmos*, giving at least passing consideration to some of the beliefs held among the Ancient Greeks. But even then, his focus is primarily on Greek philosophical and religious ideas such as Platonic philosophy that contribute to the development of Christianity.

Admittedly, the book you're currently reading is also guilty of a Christianity-centered focus when it comes to its focus on religion. I will, from time to time, have a few words to say about Greek religious beliefs or Islam, for example, but I will have far more to say about Christianity. There's a reason for that, though. This book challenges the modern Western view of a conflict between religion and science, and Christianity is the dominant religion in the West. Most American readers are likely to be Christians, and even though Europeans are likely to be less insistent about their Christianity, it is still true that most of the people of Europe are Christian or most familiar with Christianity among the major religions. Additionally, since I wish to challenge the view that religion and science have historically been at odds, it's best to stick to the religion most likely to be the focus of attack by advocates of the conflict thesis—the argument that religion and science have always been in conflict.

So if modern writers and television shows don't always pay close enough attention to what they mean when they use the term "religion" in discussions about the relationship between science and religion, what about the term "science?" Surely professional scientists such as Richard Dawkins, Carl Sagan, and Neil deGrasse Tyson have thought about this term, and the complexities inherent in what science is and how it works—right? To get at that question in a manageable way, let's use Neil deGrasse Tyson's understanding of science as an example. Helpfully, he provides a thoughtful and clear consideration of that very topic in an essay first published in 2015 titled "What Science is—and How it Works." This essay is both readable and sophisticated, while also readily available online. In this essay, Tyson states "science discovers objective truths," and reinforces one element of the importance of the social nature of science—the process whereby scientists review and support or refute one another's work. The essay is a good read and a good place to start when it comes to understanding what science is, but Tyson glosses over important points and may be a bit too close to his subject to recognize certain strengths and weaknesses about how science works—points that will be useful for us to understand because they directly impact our current consideration.

To begin, it's important to note—as Tyson rightfully does—that science is a historical creation, a method of inquiry developed in the seventeenth century, which isn't always given enough thought when people consider the historical relationship between science and religion. We'll discuss that a bit more later, but for now we should note that it's a method that is far messier than Tyson makes it appear when he states that the scientific method leads scientists to "conduct experiments to test [a] hypothesis and allocate . . . confidence in proportion to the strength of . . . evidence." While it's true that science privileges ideas about the way things work and explanations about why things are as they are—a hypothesis—supported by empirical evidence that is statistically and mathematically organized, quantified, and elaborated, it's also true that what counts as evidence, an explanation, and a host of other considerations are all determined by social forces at work in the society that produced the scientist. Furthermore, as made clear from an analysis of the actual histories of scientific discoveries presented in studies such as Stephen Jenkins' *How Science Works: Evaluating Evidence in Biology and Medicine,* in spite of what science textbooks across the world teach students, it's actually fairly rare for scientists to begin with a provisionally held hypothesis as a starting point for investigation. Instead, scientists often—perhaps normally—begin with a model for the way things work or that explains an important phenomenon, then work backward, as it were, to develop hypotheses that might explain the theory the scientist is building. Additionally, scientists often develop these hypotheses to fit with data they've gathered by a sort of playing around with experiments, rather than developing experiments for the express purpose of testing a previously developed hypothesis. Furthermore, every question asked, experiment developed, and every analysis of data occurs within the social environment of the scientist's larger world, as well as the social environment of the scientist's profession.

All of that might sound as if it's meant to take science down a peg or two, but that isn't the case at all. Science is a powerful method of inquiry that leads to very sophisticated and useful explanations of phenomena and can lead to concrete technologies and methods for improving our lives. It's the social nature of science that means every bit of scientific data, every experiment, every hypothesis and theory, is provisional and only important for the consensus it can develop among scientists—or fail to develop. As Tyson points out, what is frequently misunderstood is that any given scientific experiment or study is more or less meaning-less on its own. Only when tested by other scientists, matched by other experiments, and joined with other puzzle pieces about the way things work does a scientific study attain real meaning, by generating consensus among most scientists that the data is meaningful and accurate. To make this abstract claim more concrete and simpler, a single study saying, for example, that eating bacon can increase the incidence of heart disease means very little. But when multiple studies about heart disease, its causes, the effects of saturated fats, and so forth, are all con-sidered together, the data and its analysis so provided can be very meaningful. If most scientists working in the relevant fields, who are familiar with these experiments, studies, and collections of data agree that in sum they support the premise that eating bacon can increase the likelihood that one may get heart disease, then this point is considered statistically proven—but still open to revision or rejection should future evidence arise. One should note that not all scientists have to agree—only most of them—for an idea to reach the level of a statistical truth. For a scientist, a statistical consensus among knowledgeable authorities is what constitutes truth.

The way in which scientists generate statistical truths is effec-tive over the long term for weeding out bad or poorly supported ideas. When one study suggests that eating bacon may be linked to heart disease, other scientists interested in heart disease will take note, sooner or later, and consider the study supporting this hypothesis as part of their own work, and if it turns out not to fit with their data, will call the study into question. True, not all scientists have to accept the premise under consideration for it to be seen as proven—only most of them. However, if enough scien-tists raise enough questions about the study, or present enough counter evidence, then eventually the hypothesis falls into the dustbin of history, to use a rather tired old phrase.

However, we should take note that it can take a long time for the scientific community to turn against an idea, especially if that idea fits with larger cultural viewpoints. So, while science is self-correcting, such corrections can take a very long time to come to pass and the process of testing, verification, or rejection are all socially bounded—they occur within specific social environments that have tremendous effects on the processes. To understand what this means and how it works, allow me a brief digression about an idea that was once considered scientifically valid and proven, which has since fallen into the previously mentioned dustbin, and thankfully so. That is the notion that white people are smarter or otherwise "better" than non-white peoples. As far back as the fifth century BCE, Hippocrates (c. 460–380 BCE) suggested that dark-skinned people are cowards by nature, while light-skinned people are naturally courageous. However, this was long before science coalesced into a method of inquiry in the seventeenth century, and such ideas didn't really develop into what historians of science now refer to as scientific racism until the period of colonialism.

By the seventeenth and eighteenth centuries, scientists such as Robert Boyle (1627–1691) and Carl Linnaeus (1707–1778) argued that people who differed in skin color were either a separate species from one another or different races, though there was considerable debate about whether these different races were the result of the climate and geography of the places where they lived or the result of more intrinsic differences. In the following century, reputable scientists such as the English physician Charles White (1728–1813) built on this idea and offered complex, evidence-based arguments in support of the idea that humanity is divided into distinct races, and that some races are superior to others. This idea was highly convenient, for by the time White was writing, Europeans had already conquered the Americas and the native populations had been largely eradicated or enslaved, while the ancient civilizations of India had been more or less reduced to an English colony. The period of imperialism, when white Europeans who had developed industrial processes of mass production and technologies of killing that the people of Africa, Asia, and elsewhere had not yet mastered, had begun. Given that larger social and historical context, it's no surprise that European scientists developed theories of European racial superiority allowing them to believe their conquest of so many non-white peoples was not only natural, but also a favor to these "lesser" people. Supported by the theory of scientific racism (though this is a modern term that these scientists would not have recognized) and the evidence gathered through observation and experimentation that scientists such as the French zoologist Georges Cuvier (1769–1832) and the German biologist Ernst Haeckel (1834–1919) presented, Europeans and Americans almost universally accepted it as a scientifically proven fact that white Europeans and their American counterparts were superior to non-white, non-Europeans, because this idea fit with the larger social body of knowledge.

It wasn't until after World War II, when scientific racism had become firmly associated with the horrors of the Nazis, that the scientific community began to turn away from this idea, a rejection that was due to social forces rather than the pure, unbiased pursuit of truth that Tyson writes about. True, scientists working across disciplines have now demonstrated the self-correcting nature of science by generating hypotheses, gathering data, and generating a consensus that rejects the idea of the racial superiority of one group over another—indeed, the modern consensus is that the concept of race itself lacks scientific validity. However, all of this work has been completed within a social context in which racism is now largely seen as not only unjustified, but also actually evil.

The discussion I have just provided about science is very preliminary and glosses over mountains of complexity. I'll flesh a few points out in the chapters to come, but any thorough analysis of the nature of science and what constitutes scientific truth would take us too far afield from our focus on the historical relationship between science and religion. However, we need to be mindful of this thumbnail sketch of what science is, as well as its strengths and limitations, as we move forward. There are a number of reasons why it's important to keep these things in mind, but one reason of particular importance to this current study is that the perceived relationship between science and religion scientists like Dawkins and Tyson maintain is the product of social forces that have little to do with the available evidence.

In order to understand that argument, we need to begin with a consideration of Greek and Roman antiquity. The reason why we start there is that modern people have a view of the Ancient world handed down to us by the Renaissance, which has serious implications for understanding the historical relationship between science and religion. This view is beautifully exemplified by both versions of *Cosmos*, which allude to these periods as eras of intellectual vitality and robust flourishing of scientific knowledge. According to the heroic vision of the history of science that both Carl Sagan and Neil deGrasse Tyson present, the religious mindset of the Middle Ages snuffed out this intellectual vigor. Tyson and Sagan are certainly correct to praise certain periods within the broad expanse of classical antiquity for the flourishing of intellectual thought that occurred, but as with the rest of the sweep of the history of science, it is far too simplistic to imagine that the Greeks and Romans were coming up with really great ideas and operating full-throttle in their examination of the natural world, until religious faith and orthodoxy crushed this spirit of inquiry. And of course the "faith" in question here is Christianity, for the Greeks and the Romans had their own faiths, a fact that often goes undiscussed in the popular narrative of the history of science. However, since the modern intellectuals who are writing reactively against faith are primarily concerned with what they perceive as a "Christian" attack on scientific ideas, Plato (c. 427–c. 347 BCE) and Aristotle's (384–322 BCE) belief in the reality and power of the Greek gods is given a pass. In spite of such issues, it's true that thinkers of Classical Antiquity did, over time, turn increasingly away from consideration of the natural world, but there's little reason to think that they did so because of the influence of Christianity or any other religion. Instead, as we will find in chapter 1 on "Greek Philosophy and the Natural World," the shift in focus that occurred in late Antiquity resulted from broad social and economic forces rather than religious opposition to free thought.

Moving out of antiquity, Sagan and Tyson point to a specific period, the Middle Ages, as a time when religion was most visibly opposed to science. In the 2014 version of *Cosmos*, the first episode devotes 25% of its content to a consideration of Giordano Bruno (1548–1600), whom Tyson confidently tells us was the only man on "the whole planet who envisioned an infinitely grander cosmos" than that conceived of by his contemporaries. Although this is really the early modern rather than the medieval era, the narration highlights all that the writers of the show—and presumable Neil deGrasse Tyson—saw as having gone wrong in the centuries after the collapse of Roman power in the West ushered in the Middle Ages. Tyson explains that the Catholic Church imprisoned and ultimately executed Bruno because he "couldn't keep his soaring vision of the cosmos to himself" during a time when "there was no freedom of thought." The story makes a nice set piece to demonstrate the danger of religious authority versus science. It also nicely shows why the writers of the show felt no need to deal with the Middle Ages, a period popularized as *The Age of Faith* by the authors Will (1885–1981) and Ariel Durant (1898–1981), the title they chose for their volume on medieval history in their encyclopedic series, *The Story of Civilization.* The Durants certainly weren't the first or the last to think of the Middle Ages as a time characterized primarily by religious faith, and there are good reasons for this viewpoint. However, for people who see religion as a natural oppressor of science, that means Giordano Bruno is not just a man persecuted for his beliefs, but also an example of scientific rationality crushed by the most emblematic institution of the Middle Ages, the Catholic Church. The perceived power of the Church and the faith that permeated the Middle Ages is why Carl Sagan's book version of *Cosmos* ends with a timeline of people associated with the study of astronomy, which includes numerous figures from Greek antiquity but skips over a 1,000-year gap before picking up again with figures from the early Scientific Revolution, Leonardo da Vinci (1452–1519) and Nicolaus Copernicus (1473–1543). Sagan glosses over the entire span of the Middle Ages as "a poignant lost opportunity for mankind."

   Cosmos—both versions—makes for compelling television, but is this view of the Middle Ages as an extended period when the dead hand of faith-suppressed scientific thought accurate? As we will see in chapter 2, "Studying the Natural World in an Age of Faith," there are very good reasons to answer no. Certainly it's an idea that is consistent with many Renaissance ideas about the Middle Ages, but far from being a period when men and women were so consumed by their faith that they either had no interest in the natural world or wouldn't allow themselves to propose ideas at odds with the official religious viewpoint of the world, there were actually exciting explosions of critical thought and consideration of the natural world—almost always by those who were part of the official hierarchy of the Catholic Church. There were certainly forces that kept the development of scientific knowledge to a slow pace, but those forces were economic and social rather than religious. The Catholic Church and its theology, far from repressing intellectual and scientific development, actually promoted these activities.

Beyond the Middle Ages, we find ourselves in the period when intellectuals created that name—Middle Ages or its Latinate equivalent, the Medieval period—in order to indicate that the classical traditions of the Romans and the Greeks was being reborn after an intervening period of darkness. Thus, the Renaissance (which means "rebirth") was seen to be a totally new age, and the years in the middle between the collapse of Roman power in the fifth century and this rebirth was one when not much very interesting or intellectual was happening, according to the writers of the Renaissance. These authors were nothing if not brilliant publicists, and the result has been an enduring characterization that most casual (and some not so casual) students of history accept. It's also a seminal period for any consideration of the history of science, as Renaissance developments in art and philosophy contributed to the birth of modern science, though not necessarily in the ways that Sagan and Tyson present. Carl Sagan tells us in episode 7 of *Cosmos: A Spacetime Odyssey* that it was during this period that "free inquiry was valued once again," suggesting that this valuation of free inquiry occurred because of the overthrow of a type of Platonized Christian mysticism.

However, this characterization is rather wide of the mark. There certainly was an explosion of intellectual development during the

Renaissance, but this explosion was the result of economic and social forces rather than any sort of turning away from Christian beliefs and the Platonic influences that had shaped Christianity. In fact, the very Platonism Carl Sagan points to as one of the causes of the decline of Greek intellectual traditions combined with Christian beliefs to lead Nicolaus Copernicus's (1473–1543) to a conviction that the sun should be the center of all things. Thus, as we'll explore in chapter 3," Medical Knowledge and the Study of the Heavens in the Renaissance," the Renaissance was a period in which complex cultural and economic changes influenced artistic and scholarly developments, contributing to scientific advancements—but none of this had anything to do with a turning away from Christianity.

The positive influence of Christianity is even stronger when it comes to Johannes Kepler (1571–1630), who was driven by his Christian faith (he was a devout Lutheran) to seek out God's divine plan through an examination of His creation, the universe. And philosophical beliefs similar to those held by Copernicus drove Galileo (1564–1642) to embrace Copernicus' sun-centered model of the universe. Furthermore, his infamous conviction for heresy was not simply the religiously motivated persecution of a brilliant scientist that it is so-often portrayed as. Instead, while religious beliefs played a role—a role made more complicated by the Reformation that had begun and sparked devastating wars of religion in the sixteenth century—Galileo's prosecution was the result of personal animosity toward a man who was deeply condescending to his peers and wrong on many points of scientific fact. We will explore all of these issues in chapter 4, "The Scientific Revolution in Germany and Italy," as we consider the interactions between religion and science and the contributions that Christianity made to the developing scientific tradition in German-speaking lands and Italy in the first stages of the Scientific Revolution.

Chapter 5, "The English and French Scientific Revolutions" will discuss the shift of the epicenter of scientific development from continental Europe to England, with a consideration of some of the cultural and economic reasons why this shift occurred. We will also see how those same forces gave rise to scientists such as Robert Boyle (1627–1691) and Sir Isaac Newton (1643–1727). Both of these men were deeply interested in alchemy and the philosophical principles that grounded it, which had a tremendous impact on their work. And both of these scientists were men of deep faith, a faith that was central rather than peripheral to their scientific work. Furthermore, while the belief that Newton's mechanical philosophy inspired in him a sort of rationalist rejection of an interventionist God is so common that this claim has made it into many textbooks, nothing could be further from the truth, as we'll see in chapter 5. However, his work and the Royal Society, of which he was the president from 1703–1727, were major influences on the work of eighteenth-century Enlightenment thinkers such as Voltaire (1694–1778), many of whom were openly hostile to organized religion. This has led modern scholars such as Philip A. Egan in his 2009 *Philosophy and Catholic Theology* to consider this period as one of simple intellectual rejection of religious ideas, a rejection that is often seen as driven by Newton's work. However, I will argue we can't understand the attitude of writers such as Voltaire, unless we place this attitude in the context of their reaction against an oppressive monarchy and other social forces. Therefore, we'll consider the late Scientific Revolution in England in conjunction with the Enlightenment in chapter 5.

However, it wasn't only the study of the heavens and the human body that experienced revolutionary changes as Europe made the transition to the modern world. In France, tensions had been building for generations as the social system became ever more skewed toward benefitting the nobility, and power increasingly accumulated in the hands of the king. This process reached its peak in the reign of Louis XIV (r. 1643–1715), who has been said to proclaim that "the state, it is me." Even as this concentration of power was occurring, the patronage of the wealthy allowed science to progress in France, though the makeup and rules of the French Academy of Science affected the way those interested in the natural work pursued their work, in ways that would prove quite different than what was occurring in England. As dissatisfaction turned to anger for many, France stumbled steadily toward the French Revolution, which would break out in 1789. Against this backdrop, the study of the natural world continued, as we'll see in chapter 6, "Revolution, the Changing Conceptions of Creation and Creatures, and the Rise of Fundamentalism."

Although chapter 6 covers a great deal of ground, considering developments occurring between the seventeenth and early twentieth centuries, the focus is always on understanding changing scientific views about living creatures, from static views to those of special creation that brings new species into the world, to an emerging sense that living creatures have somehow evolved over time. As chapter 6 discusses, the French intellectual, Jean-Baptiste Lamarck (1744–1829), proved highly influential in the developing understanding that living things have changed greatly over time. However, chapter 6 will follow the ways in which the epicenter of scientific development shifts from France to England yet again, and the cultural, economic, and social milieu that produced Charles Darwin (1809–1882) and allowed his theory of evolution by natural selection to gain relatively quick and widespread acceptance. Many people of faith had no problems with Darwin's theories, but of course there were some who found his ideas abhorrent. Thus, we'll consider how religiously motivated opposition to Darwin's theory developed, as well as the ways in which this opposition was part of a broader suspicion of modernity that was growing in intensity among a subset of people of faith. This opposition led to the development of Christian Fundamentalism, and a consideration of this development marks the conclusion of chapter 6.

Similarly, while Enlightenment criticism of organized religion certainly laid the groundwork for the antagonism that many intellectuals of the nineteenth century developed toward Christianity and informed the work of what is likely the most influential books in the development and transmission of the conflict thesis, John W. Draper's (1811–1882) *History of the Conflict between Religion and Science*, there were other social forces at work than a simple antagonism toward religion. Published in 1874 in the wake of massive Irish immigration into the United States, there are critical social forces that need to be examined if we are to understand why and how Draper made his argument, which I'll provide in chapter 7, "Science and Religion in Conflict?" In spite of the large number of religious leaders who spoke out in favor of Darwin's theory in the nineteenth century, men such as Cardinal John Henry Newman (1801–1890), for example, there were those such as Andrew Dickson White (1832–1918), the American historian and first president and co-founder of Cornell University, who felt science was suffering prosecution at the hands of religious people. That personal feeling, coupled with the same anti-Catholic feeling that motivated Draper drove White to become the most influential promoter of the Conflict Thesis, with the publication of his *A History of the Warfare of Science with Theology in Christendom* in 1896.

However, it's too simplistic to paint Christianity as reacting against science at the close of the nineteenth and the opening of the twentieth century. Even as White was finishing his *A History of the Warfare of Science with Theology in Christendom,* a man was being born who would show why such simplistic notions were never very accurate. This was the Belgian priest and physicist, Georges Lemaitre (1894–1961), whose work laid the foundation for what is now commonly known as the Big Bang Theory. Although this cosmological theory has become—along with the theory of evolution—a flashpoint for a certain kind of fundamentalist Christian, not only did a person of faith develop this theory, but initial resistance to it came primarily from non-religious scientists. Those such as the astronomer Fred Hoyle (1915–2001)—who created the name "the Big Bang theory" as a form of attack on Lemaitre's theory—and Albert Einstein (1879–1955) saw the theory as overly religious in its conception. Needless to say, most of these scientists followed Einstein in coming to accept what Lemaitre called the Primal Atom or Cosmic Egg theory as a model for the way the universe came into being and developed, along with most religious leaders. In 1951 Pope Pius XII (1876–1958) formally announced that there was no intrinsic conflict between the Big Bang theory and Christianity (he had made a similar pronouncement about Darwinian evolution the year before) and other major Christian groups such as the Evangelical Lutheran Church of America, the Episcopal Church USA, and other mainstream protestant organizations have stated support for the acceptability of modern evolutionary and cosmological theories.

Following chapter 7, I'll have a few things to say in my conclusion about current views regarding religion and science. Although modern proponents of the conflict thesis such as Lawrence M. Krause (1954–present) would have us believe that these two ways of understanding the world are inherently in conflict, others such as Sir Robert Boyd (1922–2004), founder of the "Research Scientists' Christian Fellowship," and the Nobel Prize winner, Charles H. Townes (1915–2015), whose 1966 work, "The Convergence of Science and Religion," have strenuously disagreed. Certainly a vocal and politically well-connected subset of religious leaders reject scientific approaches to knowledge such as evolutionary theory and modern cosmological models, but this is part of a broader concern about modernity rather than being intrinsic to a Christian worldview. That fact is apparent from the large numbers of religious leaders who have expressed support for or at the very least acceptance of modern scientific views of the development of life and the cosmos.

Anyone reading this book to its conclusion will see that the historical evidence simply doesn't support the conflict thesis. However, this book isn't intended to be a comprehensive exploration of the history or the philosophy of science, though I'll draw from each of these areas as needed. Instead, it is an effort to replace a flawed view of history in which religion and science are perceived as always at odds, with one that is more historically accurate. To that end, I will argue that religious and scientific beliefs and knowledge have reinforced one another far more often than they have come into conflict, using historical examples drawn from different eras to support and explain this argument. I make no claims for being comprehensive, as it would be impossible to write a comprehensive history of the relationship between science and religion in the West in ten times the space I've used here. Instead, each chapter considers a handful of important case studies, treated with enough depth to hit the highpoints, but distilled down into what I hope is an easy read.

Just as this book isn't an in-depth historical or philosophical examination, it's also not a work on science, for the very good reason that I'm not a scientist. I'm a historian. Therefore, the pages to come will be filled with historical examples and occasional philosophical considerations, but they will contain no more science than is necessary to give a very barebones explanation of what any given scientific idea is. And while I will discuss points of conflict, such as that between fundamentalist Christian beliefs and Darwinian evolution, I won't attempt to resolve those conflicts. That isn't because I have no opinion about these conflicts, but rather because there are plenty of other books that analyze, explain, and offer resolutions to these conflicts. This book is for those interested in understanding the historical relationship between science and religion, rather than those wishing to learn the science or the religious notions inherent within these conflicts.

Finally, I will admit to one final hope I have for this book, though I wouldn't go so far as to say it's an actual goal. A goal would make it seem as if I think I can bring this hope to fruition, which I think might be a bit too much to ask for. My hope is that by understanding the historical relationship between science and religion, more people will see that these two ways of understanding the world are not a threat to one another. Thus, maybe people of faith can be more comfortable with science, and those who favor science will be concerned less about religious faith. I doubt this modest book can bring about any large-scale revolution, but if it helps even a handful of people come to a more historically accurate understanding of the relationship between religion and science, it's a success.

CHAPTER 1

Greek Philosophy and the Natural World

A s we will discover, John William Draper's 1870 publication, *The Conflict Between Religion and Science* is a frequent touchstone when it comes to understanding the popular view of the relationship between religion and science. Alongside Andrew Dickson White's (1832–1918) two-volume *A History of the Warfare Between Religion and Science in Christendom*, published in 1896, these works not only described the attitude of many among the American and European intellectual elite, but they also promoted the viewpoint that science and religion—or more precisely for these authors, Catholicism and science—had always been in conflict, and in fact *must* conflict. Their works proved both influential and popular, not only among the aforementioned intellectual elite, but also among many middle-class Americans and beyond who wanted to be seen as well-read and articulate in their efforts to rise upwards through the ranks of society. Jointly, these two works have gone through more than eighty editions. Draper's book has been translated into languages that include Spanish, French, German, Polish, and Russian. White's has been translated into French, Italian, Swedish, and Japanese. The last edition of Draper's work was in 1970, but White's work is still in print, and presumably those who still buy and read it also accept his views. The central point in their arguments is that

Christianity "crushed" and "perverted" science, as White put it, snuffing out a pagan culture that had privileged free inquiry, which "asserted that knowledge is to be obtained only by the laborious exercise of human observation and human reason," as Draper put it. For these key promoters of the Conflict Thesis, the culture of the ancient Greeks and Romans was one in which philosophy, science, and free thought reigned supreme until the rise of Christianity, which, in the words of Draper, "was ever ready to resort to the civil power to compel obedience" to orthodox thought. This created "a stumbling block in the intellectual advancement of Europe for more than a thousand years," during which time science was crushed beneath the dead hand of theology.

Draper and White spin a good story, though it's worth noting that while White was a professional historian—he earned his M.A. in history at Yale and went on to hold a variety of academic positions, most notably as the first president of Cornell University— Draper was not. He wrote a good deal of history, but was by training a chemist, and as such had all the biases of a scientist of his day, which we will explore more fully in chapter 7. For now, we'll focus on whether or not they were right about Classical Antiquity. Were the pagans of Ancient Greece and Rome dedicated to free inquiry and making great strides in science? Were there forces other than Christianity that could have threatened this intellectual vigor? As we will see, the evidence shows pagans weren't always as dedicated to free inquiry as Draper would have us believe. Furthermore, while numerous Greek thinkers made strides in laying the foundation for procedures for understanding the natural world, we should be cautious about referring to what they did as science. We should keep in mind that those who studied the natural world were always teachers rather than what we would think of as scientists, and in most instances had few incentives to do anything resembling applied scientific work. And as for the claim that any decline in intellectual vigor occurred due to Christianity, that claim is simply false. Rather, due to internal social forces, Greek intellectual thought began to be redirected into channels other than those that focused on an exploration of the natural world long before Christianity made any impact.

In ancient Greece there were two different approaches to understanding reality, one of which privileged observation of the natural world, while the other focused on reasoning rather than observation. In many ways these two approaches were related, but it's important to understand the differences. Both began among Greek immigrants, in a region known as Thrace, which is centered on the modern borders of Greece, Turkey, and Bulgaria, and on the Ionian coast of Asia Minor. Greece itself was very resource poor, with limited good land for growing crops, so as the population of city states (Greece was not unified as a single country, but made up of self-governing cities that controlled the land immediately around the city) such as Athens or Sparta on the mainland grew, citizens emigrated to areas around the Aegean or Black Seas. The colonies formed themselves into new city states, which remained closely associated with the city states of Greece proper. The people who lived in these cities were still considered Greek. They also maintained Greek customs and values, including placing a premium on education, at least among male citizens—that is, the Greek men who owned land. Anyone interested in a career in politics, for example, understood that being well educated was a necessity, and teachers were held in high esteem. However, we need to remember that the Greek model of education was quite different than our modern one, and while many (but not all) educated Greeks saw study of the natural world as important, this study was in no way like our modern science. Instead, natural philosophy, as it was known, was simply one of the many areas of study any Greek teacher would be expected to be knowledgeable about. Furthermore, there was no social value placed on a focused study of the natural world or the development of concrete, testable new ideas about the world as is the case with modern science. Natural philosophy was the province of teachers who sometimes developed new ideas about the world, but more often simply passed along the knowledge of the philosophers who came before them. The lack of prestige associated with doing anything like applied science is one of the reasons why, over the centuries, Greek thinkers wrote many words and developed many different ideas about the world, while doing very little to develop testable notions or new technologies.

Nevertheless, it is in these areas where Greek citizens had migrated away from their homeland that we find the beginnings of Greek philosophy and what scholars like the political philosopher Eric Voegelin (1901–1985) have termed the "Greek discovery of reason." That might be too broad a claim, since the Egyptians and Babylonians—not to mention the Chinese and Indians—certainly developed sophisticated tools for the application of reason, but there's no doubt that the Greek approach to reason has had a tremendous influence on the development of Western thought. And it was those men known as the pre-Socratics, because most of them came before Socrates (c. 470 BCE- 399 BCE), who first developed the Greek approach to reasoning. Among these thinkers was an oddly eccentric and very private man by the name of Pythagoras (c. 570–c. 495 BCE). Unfortunately for us, most of Pythagoras' writings have been lost and the primary sources we have about his life were mostly written centuries after his death, though the best of these sources are likely drawing on sources closer in time to Pythagoras' life. Other sources include stories gathered about him by later writers who were fascinated with his work, which can be dismissed out of hand: he certainly wasn't the son of the god Apollo, nor did he have the mystical power of being in more than one place at the same time. However, if even a fraction of the tales about him are true, then his brilliance as well as his personal oddness can't be questioned. One thing we do know for sure about him, though, is that as far as he was concerned, any focus on the material world could only lead one astray when it came to the pursuit of knowledge.

Born to a father who was a craftsman or merchant on the island of Samos in the Aegean Sea, Pythagoras reputedly traveled widely, to Greece, Egypt, and perhaps even as far as India. Around 530 BCE he settled at Croton in Magna Graecia—the coastal areas of Southern Italy that became a home to many Greek immigrants. He established a school of sorts there, where he taught for ten years before going back to Samos. There's really no way to know precisely who taught Pythagoras, though two

Egyptians, Soches and Plato of Sechnuphis (not to be confused with the more-famous Greek philosopher also known as Plato) are both frequently mentioned. Furthermore, it's all but impossible to figure out what, exactly, Pythagoras' contributions to the history of ideas and science were, since the ideas of the Pythagoreans as a whole—Pythagoras and the generations of followers who came after him—are all muddled together. Even the most famous idea Pythagoras presumably developed, the Pythagorean theorem familiar to every student of geometry (the square of the side opposite the right angle, the hypotenuse, is equal to the sum of the squares of the other two sides), is not attributed to him until some five centuries after his death.

However, we can figure out at least the broad strokes of what he taught by examining the beliefs of his followers and what the sources say about him. It seems that Pythagoras was either wholly or partly a vegetarian, and forbade his followers to eat meat or wear wool or leather. He may have developed this position because of a belief in reincarnation, as some stories include statements he made about the souls of humans returning in the bodies of animals. One such story tells of Pythagoras intervening to stop a man from beating a dog, because he said that he recognized the voice of a dearly departed friend in the dog's howls. Pythagorean philosophy seems to have been conservative in nature, which sometimes caused problems for his followers; presumably, citizens of Croton destroyed his school and chased his followers from the city when they attempted to block the establishment of democracy in the city. Most importantly, and most well documented, is his belief that nature was rationally explainable. However, unlike other Greek thinkers who posited a physical substance underlying all of nature, Pythagoras saw the abstract principle of number as the force unifying reality.

This idea that mathematical and numerical relationships represent the fundamental order of the universe is a notion with an enduring importance. Anyone who has ever taken a college-level science class knows how important mathematics is to the study of science. In fact, science without quantification, data crunching, and mathematical elaboration simply isn't science.

However, Pythagoras and those who followed him didn't approach mathematics in the same way that a modern mathematician or scientist does. For Pythagoreans, understanding numbers was synonymous with understanding reality—the physical world around us isn't real in any fundamental sense, but is, instead a distraction from that which is real. While that may be a confusing notion to a modern reader, I invite you to keep in mind what a poor job our senses actually do at imparting knowledge about reality. When we watch the sun or the procession of stars across the night sky, for example, it certainly looks as if the Earth is at the center of the cosmos. And many, many studies show that eyewitness testimony in a court of law is actually the least reliable form of evidence, as we only "remember" what we think we saw in the past, and anything we think we saw is influenced and shaped by our biases, preconceptions, and the limitations of our sight, hearing, and other senses. Therefore, this notion that the observable world is a distraction from understanding what is truly real will prove to have enduring power, and in fact plays a role in the structure of the scientific method when it develops, as we shall see later in this book.

The Pythagoreans recognized that the mathematical relationships they found so fascinating had a direct connection to the sensible world, meaning the reality accessible to our senses. One apocryphal story tells of Pythagoras walking by a group of blacksmiths at work, marveling at the different tones their hammers made as the men pounded out metal objects on their anvils. Upon investigation, he determined that there was a direct relationship between the length and weight of the hammers and the tones they produced. We know this story is apocryphal because the relationship he presumably worked out doesn't work at all with objects like hammers—but it does work perfectly for the tones produced by a string on a musical instrument. When a string on an instrument such as a harp produces a note, the pitch can be changed by an octave by reducing the vibrating portion of the string to half its length, and can be changed again by shortening it to a third, two-thirds, or three fourths of its original length. This is a directly observable way in which there is clearly a relationship between the sensible world and numerical relationships, but for the Pythagoreans, it was the underlying mathematical principle that was truly real and important, not the actual vibrating strings. While we don't know if it was Pythagoras himself who discovered the relationship between musical notes and mathematics, it may well have been. Regardless of who discovered it, though, Pythagoreans remained fascinated with the mathematical complexities of music. This fascination included the development of a notion that turned out to be false, yet was influential for many, many centuries; given that musical notes from a stringed instrument depend on the rapidity of the vibration of the string, Pythagoras (or one of his followers) posited that since the planets move at different speeds, they must also produce different sounds as they move. The resulting sound of the motions of all the heavenly bodies would be a beautiful symphony—the so-called harmony of the spheres. This idea would prove to be extremely attractive for a long time—Johannes Kepler (1571–1630), about whom we'll hear much more later, wrote a study of this supposed phenomenon.

However, in spite of the contributions that Pythagoras and the Pythagoreans would make to mathematics, he tends to get a bad rap by scientists. Carl Sagan, in episode 7, "The Backbone of Night," of the original *Cosmos* series, has a few things to say about Pythagoras and his followers. In a sad and somber tone, Sagan states that "ordinary people were to be kept ignorant" about the work they were doing. He goes on to note that "instead of wanting everyone to share and know of their discoveries, they suppressed the square root of two and the dodecahedron," though it's not clear how many people were clamoring to learn about the dodecahedron. In some ways, the manner in which the 2014 version of *Cosmos* treats Pythagoras is even worse, since it ignores him entirely. This approach is more or less consistent with popularly held histories of science. For example, both Draper and White ignore Pythagoras entirely, except for a single passing reference in each of their books to the Pythagorean belief that the sun was at the center of the cosmos, with all other things orbiting around it.

Why do scientists and traditional histories of science tend to speak ill of Pythagoras? Quite simply put, it's because he doesn't fit into the narrative that so many people believe to be true. Pythagoras was utterly convinced that reality is governed by reason and organized rationally, but this belief was one that was religious rather than empirical. Pythagoreanism was something of a cult, with members who swore oaths by and recognized one another by their knowledge of the tetrachys, a triangular figure made up of four rows that add up to the supposedly perfect number of ten. Their beliefs were religious, and as such—as is true of most religions—they had no interest in sharing the intimate details of their faith with unbelievers. Furthermore, in contrast to the narrative that pagans promoted free inquiry, Pythagoreans suppressed or at least refused to talk about the fact that the sides of the isosceles right triangle could not be expressed in a convenient ratio to the hypotenuse, in terms of integral numbers. This discovery was very uncomfortable to Pythagoreans, for it demonstrated that there are some magnitudes that aren't commensurate with one another. A modern mathematician solves this problem with irrational numbers, but for the Pythagoreans, to admit the possibility of irrational numbers would have been to toss out the very rationality of the universe that provided the basis for their entire belief system. Of course they attempted to suppress such knowledge, for they didn't think of it as knowledge it all—they saw it as obviously false, and believed that there *must* be a rational answer that they simply hadn't found yet.

However, it's a mistake to exclude Pythagoras from the history of science, for as we shall see, he had a great influence on Plato, whose similar rejection of the fundamental reality of the sensible world had a significant influence on the development of science as we know it. Additionally, his central insight—that reality is mathematically describable—is *the* crucial element of modern science. The rejection or denigration of Pythagoras and his followers has more to tell us about the skewed view of the relationship between religion and science that so many Western people hold than anything about history. Contrary to the view of religion as repressive and alogical, Pythagoras strove to understand reality and devoted his life to the search for truth through rational analysis *because* of his religious beliefs. As we will see, this religious motivation to understand reality that leads to the development of scientific beliefs has been the norm rather than the exception in the history of science.

However, Pythagoras' metaphysical approach was only one thread of thought in the Greek intellectual tradition. Another line of thought is represented by Democritus (c. 460–370 BCE), who is sometimes known as the "laughing philosopher," and thus is often portrayed in paintings as a laughing man. This nickname comes from his habit of mocking and questioning people's beliefs, particularly in regard to ethics, not so much because he was an iconoclast or an ass, but more because he wanted to provoke his fellow citizens to think about why they held their beliefs, and why they did the things they did. Therefore, he regularly questioned and challenged people about their beliefs in a manner similar to his contemporary Socrates, even though Democritus is known as a pre-Socratic. However, according to numerous stories he was also a generally jovial and happy-go-lucky person, so he may well have enjoyed a good laugh on a regular basis.

According to the sources, Democritus' father was quite wealthy, so rich in fact that he entertained the Persian king, Xerxes I (r. 486–465 BCE), and his army when he marched through Thrace on his way to the invasion of Greece early in his reign. As a reward for this service, King Xerxes left several Magi to instruct the local population of Democritus' home city of Abdera, in Thrace—a region centered on the modern borders of Bulgaria, Greece, and Turkey. These Magi were Zoroastrian priests; Zoroastrianism is the world's oldest monotheistic religion, built around the idea that all of creation emanates from the deity Ahura Mazda. Many elements of Zoroastrianism, from its monotheism to its emphasis on free will, may have influenced the development of religions such as Judaism, but what is most relevant here is that the Magi were highly educated in many areas, including the study of astronomy, which also encompassed what we would today call astrology. It is from these Magi who stayed behind in his home city of Abdera that Democritus learned about astronomy. It's also likely that his association with them inflamed his desire to learn more about the outside world, for he soon set out on a wide-ranging international tour in order to learn all that he could. He lived in Egypt for years while he studied mathematics, before going on to study with the Chaldeans in the near east, and he may have traveled as far as India. He also traveled throughout Greece, where he learned from many great teachers and bought manuscript copies of the writings of the most famous Greek philosophers.

While in Greece, he visited Athens and it's possible that he met Socrates and Hippocrates (c. 460–370 BCE), who is sometimes known as the "father of modern medicine." However, it seems that he didn't study with the greatest of the Greek philosophers of his day, Socrates. One charming story relates that when Democritus met the famous teacher, he was too shy to even say hello. More influential than any meeting he may have had with these intellectuals were the writings of Anaxagoras (c. 510–c. 428 BCE) and Leucippus (fifth century BCE). Very little is known about these men today, and as far as Leucippus is concerned there were fiery arguments among nineteenth century German classicists about whether or not he even existed. Today, most scholars accept the historical reality of his existence, but it's very difficult to separate his teachings from those of Democritus, since none of his writings have survived and what we know of him is primarily through the writings of Democritus, who may have been his pupil. This situation isn't that uncommon among ancient writers—as we saw, the only things we know (or think we know) about Pythagoras is what his later followers and others said about him, and the same is true of Socrates, whom we only know through the writings of Plato and a few other references, though in the case of Socrates he apparently never authored anything in the first place. What we do know about Leucippus is that he posited that all things are made from miniscule, imperishable, and indivisible elements known as atoms (the word "atom" means "indivisible" in Greek). As for Anaxagoras, he argued that the world is made up of different combinations of primary, imperishable substances in which the various substances of the world occur not because of a preponderance of any one primary element, but rather it is the mixtures of these elements that creates the substances visible to us. Everything is ordered by a force known as *Nous* (Mind), which separated out and combined the substances that were originally blended into a single homogenous mass. The really important part of Anaxagoras' writings, though, from the perspective of their influence on Democritus, is that he argued that all natural phenomena could be explained through interactions of physical substances, rather than appealing directly to supernatural explanations.

Democritus eventually made his way back to Abdera, where he took up teaching and giving public lectures in order to support himself. Armed with the teachings of Anaxagoras and Leucippus, as well as the knowledge he'd gained from various teachers and personal observations made during his travels, he expanded on what he'd learned. His lectures and later writings include discussions of ethics—though he either lacked a systematic approach to ethics, or nothing he said that was systematic has survived—aesthetics, and the subject that he has become most famous for, atomic theory. We should keep in mind that his understanding of an atom wasn't the same as the modern conception. For Democritus, atoms were eternal, having always been in motion, and continuing in motion forever. There are both an infinite number and an infinite variety of atoms. According to Democritus, atoms are inert, filling space but not reacting to one another in any fashion, rather than the modern model in which atoms are held together or pushed apart by forces of attraction or repulsion, Democritus envisioned atoms as being held together with hooks and eyes or ball and joint style of connections. He also believed they existed within empty space, which he termed vacuum, which means "not being" in Greek. For Democritus, observable events—even those involving somewhat abstract elements such as smell or taste—are the result of the mechanical interactions of atoms with one another, or with the collections of atoms that make up observable physical objects.

Democritus' atomic theory is rather brilliant, and since Democritus' atoms were unobservable due to their smallness, students in the classroom normally find themselves stumped when trying to figure out how he could have come up with the notion. However, there's nothing magical about how he came by his ideas—it's a simple matter of inductive reasoning. Democritus wrote of a city where travelers entering and leaving would rub the head of a bronze statue for luck. Although it wasn't possible to observe any portion of the statue being rubbed off, over time the statue had been worn down to the point where its features were mostly gone. Therefore, as Democritus correctly reasoned, each time someone rubbed the statue, they must be rubbing bits far too small to be seen by the naked eye off in the process. Democritus posited that these bits too small to see were atoms. Other proponents of atomic theory, such as the Roman poet, Lucretius (c. 99–c. 55 BCE), would refer to observable phenomena such as the slow erosion of rock by the actions of water as evidence that all things are made of atoms too small to be observed.

Scientists tend to see Democritus as something of a hero, a man who allowed inductive reasoning—the ability to come up with general principles by examining particular instances— to trump superstition. In the seventh-episode of the original *Cosmos,* "The Backbone of Night," Carl Sagan called his ideas "subtle and elegant" and "fundamentally right," which is what counts for high praise from a scientist. Perhaps more telling, though, is the fact that Sagan and his wife Ann Druyan named their son after Democritus, although perhaps thankfully for him it is his middle name.

In many ways it makes sense for a scientist to hold Democritus in such high esteem. After all, his ideas were brilliant and well ahead of their time, and they do represent a triumph of inductive reasoning—the kind of reasoning that makes science, well, science. We shouldn't forget, though, that it's a mistake to call Democritus a scientist, or even "the father of modern science," a title attached to his name in his entry in the *Encyclopedia of Literature and Science.* For one thing, while Democritus did look at the material world around him for examples and inspiration, it wouldn't have occurred to him to gather data the way a modern scientist does. As for the concept of carrying out an experiment, that idea would have been alien to him—not to mention that no experiment that would have been available to him could have offered empirical confirmation of his ideas about atoms. And while he wrote on mathematics and made contributions in the field of geometry—he was the first to correctly observe the relationship between the volume of a cylinder or prism and the respective volume of a cone or pyramid of equal-sized base and height, for example—for him mathematics represented a type of metaphysics, an examination of a higher-order of reality. With obvious exceptions such as base arithmetic and counting, for Democritus, math had little to do with understanding physical reality. All of his ideas were the result of pure reasoning divorced from any careful consideration of the physical world, much less the gathering of mathematically quantifiable data or experimentation.

It's also a questionable proposition, at the very least, to represent Democritus as an early atheist or something of that sort, as scholars such as Jonathan Barnes tend to do. While Democritus has some very interesting and often confusing things to say about the gods, we have very little basis on which to argue that he rejected their existence or importance. He discusses the gods in various places in his surviving writings, such as in his discussion of how we come to know things. As an atomist, Democritus believes that we can't really *know* things about the external world, for all that we can know about objects external to us is through the transmission of atoms from that object to us. Thus, all we can know about the world around us occurs through these fragments from which we attempt to construct meaning. Again, it's a rather sophisticated and interesting analysis, and Democritus shows considerable commitment to this principle of knowledge by following it through to how humans can know about the gods, which he describes as "films" of atoms that may or may not be immortal, it's very hard to tell from Democritus' surviving writings, but they are definitely difficult to observe, contact, or otherwise know because of their insubstantial nature. Regardless of whether or not he believed the gods to be immortal, there's no reason to believe that he didn't think of them as, at the least, very long-lived, powerful, and important. As the classicist Jacob Mackey put it, Democritus "conceives of them [the gods] as diaphanous beings of pure mind or soul, as ethereal as his thoroughgoing materialism will allow," but he definitely saw them as real beings. Modern people who see him as an atheist who overthrew superstition in order to embrace pure reason base this view at least as much on what they want to believe as they do on the evidence available.

The point about Democritus, from the standpoint of our current study about the relationship between religion and science, is that he is representative of a trend in Greek thought, one that emphasizes a focus on the material world in the quest for knowledge. Therefore, it makes sense that he would be something of a hero to modern scientists who see science as beginning in ancient Greece, even though he's not necessarily the earliest or even the most important representative of such a trend. Many would argue that Thales of Miletus (c. 624–c. 546 BCE), who among other things dramatically predicted the solar eclipse of May 28th, 585 BCE as a demonstration of the power of his materialist views of causation, should hold that honor. Democritus is merely the best known of these early pre-Socratic philosophers, as they are known, even though he was actually a contemporary of Socrates. And his thought represents only one line of reasoning in the Greek intellectual tradition, a line that isn't by any means dominant.

For the view that held greater influence, we should turn now to a student of Socrates, Plato (c. 427/28–347 BCE). Although his given name at birth was Aristocles, it is by his nickname that he is known to history. This nickname seems to have been derived from the Greek work *platos* or *platon*, meaning "broad," and may have referred to the wide-ranging nature of his thought, his broad forehead, or—most probably—to the fact that he was very broad across the shoulders. Born in Athens at the peak of the city state's wealth and influence in Greece, all Athenian citizens were expected to be well rounded and trained in both physical and mental arts. In line with this heritage, Plato was an accomplished wrestler as well as a philosopher of unequalled brilliance, and it may well have been his wrestling coach who gave him his nickname.

Both Ariston, Plato's father, and Perictione, his mother, were from ancient and noble families, and it would have been expected for Plato to go into politics, which is one reason why he received a comprehensive education that emphasized philosophy and the tools of logical debate. Athens was a democracy—though far from a complete one— where all adult land-owning male citizens could vote not only for important representatives to serve in the *boule* (a council of 500) and the *prytany* (a council of 50), which were jointly responsible for daily administration of the city, but also important officials such as generals. Since such a premium was placed on the power of the people, the ability to make speeches structured around logical arguments was essential for anyone who wished to obtain political power. However, Plato grew up during a trying time in Athens. The Peloponnesian War started in 431 BCE, a struggle that would engulf all of Greece as Athens and its allies battled Sparta and its allies for supremacy in Greece until 404 BCE. This war lasted an entire generation, and though it was a long, gradual process, the end result was a crushing defeat for Athens. During the years of war Athens became largely impoverished and there were times of famine when the Spartans made food shipments into the city difficult. In the end, Sparta forced the Athenians to give up their beloved democracy and accept rule by a group known as The Thirty Tyrants, though it would not be long before the Athenians revolted and reestablished their democracy.

While all of these events would have had a profound impact on Plato, what was probably the most important event in his life occurred in 399 BCE. In that year, the Athenian government arrested Plato's beloved teacher, Socrates. This was a time in which Athens was struggling to regain the prominent position it once held in Greece, when the leading elite feared the morale of the people had been degraded beyond repair, and even feared that the people might turn against democracy itself. In that atmosphere Socrates' habit of questioning the most cherished ideals of Athenian society—even going so far as to praise Athens' nemesis, Sparta, from time to time—was intolerable. The politically powerful arranged to have Socrates arrested and charged with impiety—not believing in the gods of the state—and corrupting the youth. Rather than mount a defense, Socrates affirmed that he promoted the questioning of all things, including the gods and Athenian democracy. Although it's likely that it would have been an easy task for Socrates to argue for the punishment of exile from Athens, he refused to do so, and in the end he was sentenced to die—ordered to drink poisonous hemlock. Again, it would have been a simple matter for Socrates to escape Athens and flee to safety, but that option seems never to have occurred to him. Instead, he spent the afternoon and evening with his pupils, affirming that as a philosopher he knew death was nothing to fear. In truth, it may just be that at seventy-one or so, Socrates had no interest in starting over in a new city. Whatever his reasons, he drank hemlock willingly, walked around until he felt his legs begin to go numb, then laid down to die. According to Plato, Socrates spoke his last words to his student Crito: "Crito, we owe a rooster to Asclepius. Please, don't forget to pay the debt." Asclepius is the Greek god of curing illness, and it's likely that Plato meant that death is the ultimate cure, releasing the soul from its entrapment in the body.

Plato was around twenty-eight at the time of his mentor's death, and if he'd ever had any interest in the political career his noble birth entitled him to, this event cured him of it. It's clear that Plato viewed Socrates as a victim of Athenian democracy, and in his writings, Plato repeatedly expresses his disapproval of democracy. However, for our purposes that isn't the most important element in his wide-ranging philosophy, which covers such a breadth of topics and does so with such brilliance that the philosopher Alfred North Whitehead (1861–1947) has famously stated that the "safest general characterization of the European philosophical tradition is that it consists of a series of footnotes to Plato." Fortunately, we need only concern ourselves here with Plato's understanding of how we can know anything—his epistemology, to use the technical term—and where his thought fits into the historical relationship between science and religion.

Plato settled down in Athens around 387 BCE after traveling to Italy and Sicily, where it is likely that he studied with the Pythagoreans in the Greek colonies on the Italian coast. He inherited land just outside of Athens proper, which included a region known as the *Akademia,* supposedly for its association with the

Athenian hero, Akademos, who had presumably saved Athens from destruction when the twins, Castor and Pollux, had threatened to destroy the city over the abduction of their sister—the ever-troublesome Helen of Troy. On this land was a grove of olive trees sacred to Athena, the goddess of wisdom, which made it a good place to establish a school. Throughout Plato's life this school was rather informal by modern standards, with no entrance or membership fee charged and no distinction made between teachers and students. Nor was there a formal curriculum. Instead, the Academy, as it came to be known, was organized along the lines of senior and junior members, and Plato as well as his closest associates would pose philosophical problems that its members would then debate while attempting to arrive at solutions. There were some lectures, but most instruction was done through dialectic, as senior members debated with juniors, who also debated with one another. Some of the questions Plato posed focused on natural philosophy, such as the one which asked students to compete to come up with the simplest solution for the observable irregularities of planetary motion. However, by and large the Academy and its students focused on the pursuit of philosophical knowledge pursued through dialectical debate, a logical form of argument in which a position is presented, counterpoints are made against that position, and then the issue at question is resolved, rather than anything with concrete application—though those with political aspirations were drawn to the school in order to sharpen their debating skill and gain the education expected of a Greek politician. The Academy remained in continuous operation for over 300 years, until 86 BCE when the Roman general Sulla (c. 138 BCE-78 BCE) cut down the sacred grove in order to use its wood for the construction of siege engines, which he then used to batter Athens into submission.

Many of Plato's books probably began as lectures or discussions at the Academy, and we can understand both his epistemology and the relationship between religion and science in his thought by examining elements of a most remarkable book titled *The Republic*. Plato wrote this book as a dialogue, a discussion between two or more characters, with Socrates (who is a frequent character in Plato's books) conversing with a variety of Athenians and foreigners. This use of dialogue was the norm for Plato, and the discussions often take the form of a dialectic. Plato is one of the few early philosophers whose works likely survive in their entirety, representing thirty-five dialogues and thirteen letters probably authored by him. In *The Republic,* Socrates discusses the nature of justice and the question of whether or not a just man is happier than an unjust one. The result is a consideration of various forms of government as well as what the ideal form of government might be, which Plato (through Socrates) argues would be a society governed by an elite made up of the wisest members of society, who would be philosopher kings. This discussion covers a great deal of ground and includes many allegories, one of which we will be focusing on here: the allegory of the cave.

The allegory of the cave is rather famous, and it's likely that many readers will be familiar with it. In the allegory, Socrates—who, as usual, is presenting Plato's position—speaks to Plato's brother, Glaucon. Socrates begins:

> let me show in a figure [meaning an allegory or analogy] how far our nature is enlightened or unenlightened: Behold! human beings living in an underground den, which has a mouth open towards the light and reaching all along the den; here they have been from their childhood, and have their legs and necks chained so that they cannot move, and can only see before them, being prevented by the chains from turning round their heads.

He goes on to paint a brilliant picture in words of the lives these prisoners in the cave might lead, and how it would affect them. He describes a fire lit at the prisoners' backs, and considers what the prisoners would think of the shadows cast by figures walking back and forth in front of the fire, often carrying objects. The figures walking between the prisoners and the fire would cast shadows on the wall the prisoners are forced by their chains to stare at. Socrates then asks Glaucon: "And if they could talk to one another, don't you think they'd suppose that the names they used applied to the things they see passing before them?" In other words, if the prisoners spoke to one another about the shadow of a book projected on the wall in front of them, they might call that shadow a book—but they wouldn't be discussing a book at all, instead they'd be discussing only the book's shadow. However, they would have no way of knowing that, for these shadows are the only reality they'd ever known.

Socrates concludes the allegory by considering what would happen if one prisoner were to be freed and forced out of the cave and into the daylight, where he would at first be too dazzled by the light of day to see anything at all. Once his eyes became accustomed to the light, he would for the first time see the real things that previously he'd only seen shadowed against the wall. At first, of course, he wouldn't believe his own eyes, but eventually the former prisoner would rush down to tell his friends that everything they'd ever seen had been a mere shadow of reality. How would they respond? Plato states, through Socrates, that they would kill the person who tried to tell them the truth about reality—which is precisely what happened to Socrates, as far as Plato was concerned. The Athenians had killed him for forcing them to recognize that what they thought was true was only a mirage, and for pointing out reality to them.

This allegory tells almost everything we need to know about the central concepts around which Plato constructed his philosophy. Like the Pythagoreans who influenced him—Plato's letters mention two prominent Pythagoreans, one of whom was by Socrates' bedside when he died and thus would have been Plato's fellow student, in addition to influences from studying with the Pythagoreans in Magna Graecia—Plato was strictly in the camp that privileged reasoning over observation in order to develop knowledge. Plato argued that the sensible world around us is simply a reflection of—or a shadow—of reality. Every beautiful thing we see is actually a pale reflection of what he referred to as the Form (or Idea) of Beauty, which is the perfect exemplar of beauty existing beyond the realm of the senses. In his early dialogues, Plato presents a handful of Forms, such as Beauty and Justice, as the only ones to exist, but later he seems to suggest that there is a Form for everything with which we come into contact. So every chair, for example, that we see or every song that we hear is a reflection of the Form of Chair or Song. Assuming that this position represents his mature philosophical system, certain Forms such as Beauty and Justice would be considered higher than those of individual objects. It's highly likely that Pythagorean influence led Plato to be so suspicious of the sensible world, and a near certainty that this influence is why he valued mathematics so highly, seeing numbers as an intermediary between the sensible world and the Reality of the Forms.

The key point here is that Plato believed that any knowledge we derive from an examination of the sensible world would be deficient at best. Everything around us participates in the Reality of the Forms, but are not in and of themselves real in any meaningful way. At first glance this seems to be a position that stands in stark contrast to the approach necessary for a scientist. However, Plato's ideas deserve a much more prominent place in the history of science than scientists sometimes allow. Turning again to Carl Sagan, in episode 7 of his series of *Cosmos*, he stated that "Plato's followers succeeded in extinguishing the light of science and experiment that had been kindled by Democritus and the other Ionians [by which he means the pre-Socratics, though really the only time he ever speaks approvingly of this group is when he speaks of Democritus and his followers]. Plato's unease with the world as revealed by our senses was to dominate and stifle Western philosophy." However, this quote reveals more about the limitations of an astrophysicist analyzing history than it does about the history of science—or even how science works and how it has developed. As we will see when we consider the events of the Scientific Revolution in Germany and Italy in chapter 4, one of the major stumbling blocks to the development of the scientific method was that the results it produced so often opposed our common-sense understanding of the world. To name only one example, everything our senses tell us supports the idea that the Earth is unmoving and that it, not the Sun, is at the center of the Solar System. Furthermore, the men who developed what we now call the scientific method were all influenced by Platonic thought, as I will demonstrate later.

So why does Carl Sagan have such a low opinion of Plato? While he says that it's because Plato and the Pythagoreans were "suppressors of knowledge, advocates of slavery, and of epistemic secrecy," that's hardly a claim matching the evidence or representing the importance of historical context. Yes, Plato advocated slavery, but all the Greeks did. Their entire society was built on slavery, and there is no reason to think that Sagan's hero Democritus would have been any more critical of the institution than later philosophers were. And the idea that Plato suppressed knowledge and advocated "epistemic secrecy"—a fancy way of saying that he believed ideas should be kept hidden away—is ridiculous. True, Plato didn't think his ideas could be understood by everyone, but that makes him an academic elitist at worst. So if the charges Sagan makes about Plato don't carry weight, why was he so dismissive of Plato? It's likely there are two reasons. Plato himself seems to have had a strongly religious outlook, after all his entire system of thought was otherworldly in the sense that he didn't see the physical world around us as real, which is why Sagan dismissively refers to Plato as a "mystic." But rather than that being a limitation, it was a strength, for it drove him to question, to analyze, and to reject the easy answers offered by his senses about how the world works, and instead to strive to understand a higher order of truth. That zeal to understand a truth beyond what the sensible world presents to us at the least borders on religiosity. But he has other beliefs that are more strictly religious; in his early dialogues he asserted that some people, such as Socrates and poets, are the beneficiaries of divine inspiration, while he also affirms that the common Greek practice of divination could allow people to know the will of the gods. Furthermore, in several of the early dialogues Plato states that while no one can truly know what happens after death, it is reasonable to believe that souls may be rewarded in some form of afterlife. Overall, Plato seems to have been a strong believer in the conventional religious beliefs of his day, which no doubt would grate on someone with a low opinion of religion. Additionally, Plato felt religion played a very important role in society; he argued that those who denied the existence of the gods and their involvement in human affairs should be sentenced to solitary confinement or even executed in extreme cases. So much for the idea that the pagan Greeks were naturally more tolerant than Christians. However, the strongest reason for Sagan's distaste for Plato is because Christian thinkers later adopted Platonism. As Sagan states, the views of the pre-Socratics, and by this he really means the views of Democritus and his followers

> were suppressed, ridiculed and forgotten by the Platonists
> and by the Christians who adopted much of the philosophy
> of Plato. Finally, after a long, mystical sleep in which the
> tools of scientific inquiry lay moldering the Ionian approach
> was rediscovered. The Western world reawakened.

Wow. Powerful stuff. Almost completely inaccurate, but powerful.

To begin to understand why it's wrong to imagine that a triumph of Platonism led to the suppression of "Ionian" pre-Socratic thought, the first thing I would point out is that there is absolutely no evidence to support the idea that Platonists destroyed the works of Democritus and his followers. The sad fact of the matter is that most of the works the ancient Greeks produced have been destroyed. Over the centuries, warfare has washed across the Greek peninsula again and again. Outsiders such as the Macedonians, Romans, and the Ottoman Turks have invaded and dominated Greece for most of its history and cities such as Athens experienced the horrors of destructive conquest and steep economic decline that led to the loss of libraries and manuscripts as the funds to protect them simply ceased to exist. Furthermore, philosophical trends come and go, and in the past, it was the norm for vellum (a writing material made from animal skin), which was expensive and difficult to produce, to be scraped clean and reused if the text on it was seen to be not worth keeping. Therefore, any intentional destruction of the works of Democritus was likely the result of scribes who needed to make a copy of something deemed more important than the works of some outdated and long-dead philosopher. All of these forces could come together to wipe out the works of any given author without suggesting some sort of conspiracy was at work.

A prime example of an author whose works would come to be completely lost in the fashion described above is Aristotle (384 to 322 BCE). It's likely that he was an excellent literary stylist, since the Roman statesman Cicero (106–7 BCE), who was fluent in Greek, described Aristotle's writing as a "river of gold." However, all we have left of his body of thought are works made up of lecture outlines or study aids that he made for himself or his students, or that students made while listening to him lecture. Over the centuries these have been compiled and analyzed by various commentators on his thought, so it's likely that we have a more or less complete understanding of Aristotle's wide-ranging philosophy. However, not a single one of the thirty-one works that bear Aristotle's name is actually one that he produced for public consumption. And it shows—his philosophy is certainly interesting and impressive, but the style of his works is difficult and off-putting to put it mildly.

Aristotle was born in Stagira, a no-longer-existing seaport town in Thrace that was on the northeastern fringe of Greece. His father Nichomachus (flourished 375 BCE) was a physician who most likely gave his son an introduction to natural philosophy and the study of nature. Nichomachus was also the court physician to King Amyntas of Macedonia (393/92–370 BCE), an association that would later prove important for Aristotle. Nichomachus died while Aristotle was still a boy, and a man about whom we know little other than his name, Proxenus, acted as Aristotle's guardian until the young man was seventeen. At this point Aristotle traveled to Athens to complete his formal education, where he would study with Plato at the academy for two decades. While still a student he began to lecture, especially on rhetoric, and at the time of Plato's death in 347 BCE he had already proven himself capable enough that he should have been a natural heir to lead the Academy. However, by this point Aristotle's philosophical approach and positions had shifted far from those of his mentor, so Plato's nephew Speusippus (c. 408–339/38 BCE) assumed the leadership of the Academy. Helped along in his decision by rising anti-Macedonian sentiment in Athens, Aristotle accepted the invitation of his friend Hermeas, the ruler of Atarneus and Assos from 351–341 BCE in Mysia, which is in modern-day Turkey.

Aristotle stayed there for three years, during which time he married the ruler's niece, but when the Persians captured and then tortured Hermeas to death, Aristotle went on to Mytilene on the island of Lesbos, off the coast of Asia Minor. While there he made extensive observations of plant and animal life—particularly the fish and other sea life on the coasts—until King Philip II (reigned 359–336 BCE) of Macedon invited him to come to court in order to tutor his son, Alexander (356–323 BCE), who would become King Alexander III—commonly known as Alexander the Great—in 336 BCE. However, it's important to note that these observations didn't include anything like modern experiments. To carry out an experiment in the fashion of a modern scientist would be to engage in the kind of low-status, manual activity that slaves or the poor did. No one of Aristotle's background would have contemplated such a thing.

Tutoring Alexander was a tremendous opportunity for Aristotle. Although many of the stories of how King Philip treated Aristotle are undoubtedly exaggerated, there's no doubt that he was held in high esteem and given many benefits while at the court. Philip wanted far more than to be the king of a poor land populated largely by shepherds, where the people spoke Greek but were seen as barbarians by their more sophisticated Greek neighbors to the south. Philip envisioned himself as a ruler of all the Greek-speaking people, whom he would weld into an empire capable of threatening the mighty Persians to his east. Therefore he needed his son, Alexander, to have the type of education that would allow him to garner the respect of the Greeks whom Philip hoped to rule. To keep Aristotle happy, Philip provided him with support by assigning slaves to bring required samples of animal and plant life from across near and far.

Upon the death of King Philip IV, Alexander was far too busy conquering most of the known world to spend time with his old tutor, so Aristotle returned to Athens by 335 BCE. The Academy was flourishing under the guiding hand of a Platonist named Xenocrates (c. 396/5–314/3 BCE), so Aristotle founded his own school, the Lyceum, so named because it was a temple to Apollo Lyceus (Apollo the Wolf God). Aristotle taught there for thirteen years until he was driven to flee Athens in 323 BCE by rising anti-Macedonian sentiment that led to the overthrow of Macedonian rule over Athens. Stating that he would "not allow the Athenians to sin twice against philosophy" by murdering him as they had Socrates, Aristotle fled to the Greek island of Euboea where he died a year later of natural causes. The Lyceum would continue to function as a school until 86 BCE when it too would be destroyed by the Roman general Sulla.

While teaching at the Lyceum Aristotle was at his most productive, writing some 200 works that dealt with practically every area of knowledge besides mathematics. Providing the first systematic philosophical system the Western world had seen, he also systematized logic, and the result would dominate Western education until the nineteenth century. Although he agreed with his mentor, Plato, that the world was rationally ordered, he disagreed as to how that order functioned or how to understand it. Whereas Plato had believed that one must understand the theoretical rationality of the ideal order that the physical world reflects—going from the general to the specific in his approach—Aristotle was exactly the opposite. For him, examination of the individual instances of concrete things in the world provides the basis for gaining knowledge. For while Plato argued that the things we see around us are only pale reflections of the Forms that exist beyond the sensible world, Aristotle argued that the Forms exist only within the individual things they give shape and function to in the world. Therefore, in order to understand the Form of "chairness" that structures all individual chairs, we should use our senses to gather information about individual chairs, then apply logical reasoning to that information in able to understand what it means for a thing to be a chair.

Aristotle used this approach for everything, whether he was writing about politics—where he sent out students to gather information about the way individual political systems functioned—to biology, astronomy, psychology, drama, aesthetics, or any of the dozens of other subjects he wrote about. Therefore, he was firmly in the Greek tradition of those who prized observation of the natural world in order to gain knowledge. However, just because he agreed with Democritus about the importance of observation, that doesn't mean he agreed with the earlier thinker about much of anything else. While Aristotle praised Democritus for arguing in a way appropriate to natural philosophy, he utterly rejected Democritus' conception that all things are made up of atoms. Instead, in a view that would come to dominate Western thought from the Middle Ages through the Scientific Revolution, Aristotle posited that all things are made up of an admixture of four substances: air, earth, water, and fire. According to Aristotle, every substance has a natural place in the world. Thus, why and how things move is based on the specific ratio of substances making that thing up. Therefore, a rock falls faster than a feather, say, because the rock has more "earth" in its makeup, and the natural place to which the element seeks to return is the center of the Earth, and the feather has more air but some earth in its makeup; the earth draws it toward the ground, while the air reduces the strength of that draw. Therefore, on Earth heavier substances have more earth than lighter subjects, making heavier objects fall faster than lighter. However, this model of physics only applies to things beneath the orbit of the moon. Objects at the level of the moon or higher are made up entirely of a perfect, fifth element, later to become known as quintessence (quintessence is literally made up of the two Latin terms for "fifth element"), which only moves in the perfect geometric shape, the circle. Thus, physics can only describe actions on the Earth—celestial objects move according to a very different mechanical system.

Aristotle holds an interesting position in the history of the relationship between science and religion. In a direct rejection of Plato's (who for once agreed with Democritus) belief that movement was eternal and had always existed, Aristotle argued from observation that all movement had to start somewhere. Given that he had no way of observing how objects move in a vacuum—which he didn't believe could exist anyway, in another rebuke to Democritus—this idea made perfect sense. After all, everything Aristotle had ever observed in motion, had moved because of some action on it. However, it also presented a problem because he couldn't accept an infinite regression of causes. In other words, every mover he'd ever witnessed had to have a mover that affected it, which then had to have an affecting mover, and so forth. The question that arose, then, was where or how did motion start? For that reason, Aristotle posited the existence of a Prime Mover, an eternal and imperishable entity that represented the starting point of all motion and change in the natural world. This Prime Mover is perfect, meaning that it is fully complete and incapable of change—after all, any change in a perfect being would mean a movement toward less perfection. Furthermore, there is only one action that is suitable for such a perfect being, the highest form of action of all—thought. But what does it think *about*? As Aristotle puts in his *Metaphysics*, thinking about nothing would hardly be worthy of such a perfect entity, but so would thinking about something that is imperfect. Therefore, "it thinks of that which is most divine and precious, and it does not change; for change would be change for the worse." But what is the thing that is "most divine and precious?" Well, as Aristotle explains: "it must be of itself that the divine thought thinks (since it is the most excellent of things), and its thinking is a thinking on thinking." In other words, the Prime Mover is the ultimate navel gazer, regarding and aware of only its own thoughts, for all of eternity. Still, if we remove this bit of extreme egocentrism, what we have is a perfect and eternal entity that is responsible for beginning all motion and starting all change in the natural world. That sounds a lot like the Christian concept of God, and as we'll see in chapter 2, medieval Christians very much agreed.

However, Aristotle differed from these later Christians who would find Aristotle's thought to be so useful and attractive in a fundamental way. He would have scoffed at the idea of worshipping the Prime Mover, for why worship an entity that is only aware of itself? But that doesn't mean he scoffed at the idea of worship. He did have negative things to say about those he called "the school of Hesiod," referring to the Greek poet who lived around 700 BCE and wrote about the gods. Furthermore, he writes in *Metaphysics*:

> From old—and indeed extremely ancient—times there has been handed down to our later age intimations of a mythical character . . . [with] details . . . subsequently added in

the manner of myth. Their purpose was the persuasion of
the masses and general legislative and political expediency.
For instance, the myths tell us that these gods are anthro-
pomorphic or resemble some of the other animals and give
us other, comparable extrapolations of the basic picture.

Thus, we need not take discussions of the gods' human or
animal-like characteristics seriously—but that doesn't mean we
shouldn't take the gods or religious worship seriously. Instead, he
says that while people should ignore the additions that had been
made over time, the central idea of the existence and importance
of the gods was a divinely inspired notion that had been passed
down throughout the ages. In the end, Aristotle's religious ideas
were more or less in line with the Greeks of his day.

Aristotle's reputation in regards to the history of science is
subject to wild swings. The evolutionary biologist Armand Marie
Leroi's (1964–present) position about Aristotle is clear in the title
of his book, *The Lagoon: How Aristotle Invented Science*, an attitude that has a lengthy pedigree—the late nineteenth and early
twentieth century author Elbert Hubbard (1856–1915) confidently called Aristotle "the world's first scientist . . . who sought
to sift the false from the true." Similarly, John W. Draper stated
that "the inductive philosophy thus established by Aristotle is a
method of great power. To it all the modern advances in science
are due." However, Carl Sagan's 1980 version of *Cosmos* presents
a dismissive view of Aristotle as just another advocate of slavery,
while the 2014 version of the show sees no need to mention him at
all. Perhaps that's because of Neil deGrasse Tyson's own dismissive attitude toward Aristotle. While Tyson allows that Aristotle
made some "legitimate, albeit simple, observations of the natural
world," as he states in his 2007 work, *Death by Black Hole: And
Other Cosmic Quandaries*, Tyson primarily presents Aristotle as
a peddler of ideas that were "badly mistaken," which could have
been refuted by experiments so basic even a child could carry
them out. But most damning, so far as Tyson is concerned, is
that "Aristotle's teachings were later adopted into the doctrine
of the Catholic Church. And through the Church's power and

influence" became enforced as unchallengeable dogma. While it seems hardly fair to judge Aristotle by the ways in which people who lived long after he died used his ideas—not to mention the historical problems inherent in Tyson's views of the Middle Ages, a period we'll explore in chapter 2—it's interesting to note how he pairs a dismissive attitude toward Aristotle with references to what Tyson sees as the injustices of medieval Aristotelianism. This pairing certainly makes it appear that Tyson's dismissive attitude toward Aristotle has less to do with the Greek scholars thought and more to do with the role of Aristotle's philosophy in medieval Christianity.

So which is it—was Aristotle the first scientist, or just a bumbler fiddling about with childlike ideas? Not to be trite, but as is so often the case, the truth is somewhere in the middle. While this is not a book on the philosophy of science, it is an altogether regrettable mistake to assume that no one ever thought to drop rocks in order to determine if Aristotle's ideas were correct, and saying as much (as Tyson has on a number of occasions) betrays some mistaken views about the way science works. Aristotle had a comprehensive system that made the world rationally comprehensible not just to those Greeks who accepted his ideas, but also for many in the Near East and Europe for more than 1,000 years during the Middle Ages. As such, people saw what this system told them they *should* see, so whenever two rocks fell, people invariably saw the heavier one falling faster than the lighter. As I mentioned in the introduction, multiple studies have shown that eyewitness testimony is inherently unreliable because (among other reasons) what people "see" is filtered through their preconceptions and beliefs. Besides, it's not that easy to judge which rock falls fastest when two are dropped from a height, as Galileo (1564–1642) discovered when he was attempting to refute Aristotelian physics—which is why he developed the inclined plane experiment that many physics students still do in the classroom. People accepted Aristotle's ideas because they were logically sound and coherent, allowed people to make sense of the world, and were based on careful observation of the natural world. However, I would argue against calling him a scientist (as

I have done elsewhere). While Aristotle believed in the power of observation, it never would have occurred to him to conduct an experiment –a manual activity unworthy of a high-status Greek man—or gather data, much less to apply mathematics to a study of natural phenomenon. One thing Aristotle agreed with his master, Plato, on, was that mathematics is a metaphysical pursuit, largely divorced from the mundane world of nature. That's why Aristotle wrote on everything *except* mathematics—his interest was in the observable natural world, not in the Ideal world of Forms that drew Plato's attention. So rather than a scientist, he was exactly what the ancient Greeks and medieval Europeans called him; he was a natural philosopher. However, he was a very important natural philosopher for the history of science, because his emphasis on observation, the rigor of his logical system, and the ideas about the natural world that he bequeathed to the world all laid important building blocks for the later discovery of the scientific method.

This book is not, however, a study of the history of science. Our focus is on the ways in which science and religion have interacted in Western history, so we will part company with Aristotle for the time being. It was necessary to lay out some of his ideas because in the next chapters we will see him come up repeatedly. But what happened after Aristotle? Did Greek natural philosophy decline because of the "mystical" views of the Platonists, or because of the influence of religious beliefs? In a word, no. First of all, Platonism remained only one thread in the Greek intellectual tradition, and many Greek thinkers continued to argue for the primacy of observation over abstract reasoning. Furthermore, many of those who were most interested in reasoning made important contributions to the development of natural philosophy. On the side of those who favored observation as the starting point for knowledge, no group was more important than those who revived atomism, beginning with Epicurus (340–270 BCE). Going even further than Democritus, Epicurus firmly stated that the gods have no influence over human life, and argued that the point of human life was to establish tranquility within oneself by limiting personal desires and banishing an unreasonable fear of the gods and death. On the side of reasoning as the starting point knowledge, we can find the mathematician Euclid (fl. c. 300 BCE), whose work *Elements* systematized geometry, and Archimedes of Syracuse (c. 287–212 BCE), though he also did work with applied mathematics.

As late as the second century CE, Greek thinkers continued to make advancements in the field of natural philosophy, and did so in ways that showed a solid interest both in the natural world and in observation. Claudius Ptolemy (c. 100–c. 170) was from Alexandria in Egypt, but thanks to Alexander the Great's conquests, the Egyptian elites were Greek in language, culture, and heritage. Ptolemy was particularly interested in astronomy and is most famous for a work that comes to be known by an Arabic title, the *Almagest* (meaning, "the greatest"), which compiled the work of past Greek astronomers paired with Ptolemy's original contributions to systematize a model of a geocentric (meaning Earth-centered) model of the cosmos. Ptolemy thought of the heavens as a realm of perfection made of a perfect fifth element (quintessence), in line with Aristotle, and as such he worked with ideals and mathematical abstractions. But at the same time, he wanted to create a system that matched observations of the motions of the planets, so he appealed to large bodies of observational information gathered by various astronomers. The result was a system in which the planets (including the Sun and Moon, for to the Greeks the term "planet" referred to any object that had visible motion in the heavens, relative to the so-called fixed stars) moved in perfect circles around the Earth. Each planet also had a smaller orbit around a point situated on the circular sphere of its orbit around the Earth. While it's true that this system would gain acceptance in Christian Europe in part because it matched Christian conceptions about the centrality of the Earth in creation, we should keep in mind that it also fit very well with the available observational data. There simply was no evidence until the seventeenth century to support the notion of a heliocentric (Sun-centered) solar system in which the Earth moves. Therefore, it's a mistake to focus on the acceptance of Ptolemy's system because it fit with a certain kind of religious dogma. After all, he was no Christian, and neither were the Greeks who accepted it.

They did so because it fit with observational data and the system of physics Aristotle had developed.

Likewise, Galen of Pergamum's (c. 129–200) work provides further evidence that neither a supposed dominance of Platonic thought nor religious dogma repressed Greek natural philosophical traditions that emphasized observation. Born into the Roman Empire and working for many years in Rome, Galen was a physician who was forbidden to dissect the human body—not because of any religious prohibition, but because Roman culture saw the desecration of a dead body to be deeply disrespectful. Therefore, his anatomical knowledge came from hands-on experience with patients, but more importantly on the dissection and vivisection of animals such as pigs and primates. Because of that, some of his ideas were mistaken, but he still made significant advances in areas such as anatomy and physiology. For example, he was the first to recognize the difference between venous and arterial blood, and he proved that the brain controls the muscles by means of the nerves. In spite of the limitations of his work, his theories dominated Western medicine for 1300 years, and as late as the nineteenth century Western medical students still read Galen's works.

The point of the preceding paragraphs is that there is no evidence for a suppression of the work of the atomists, or a retreat from observation and a focus on the natural world, due to the influence of Pythagorean or Platonic beliefs or because of the influence of religious beliefs—Christian or otherwise—in the Greek intellectual world. It is true that there was a decline of intellectual vigor in the Greece of late Antiquity, but that had nothing to do with religion or Platonism. Instead, it was the result of war, conquest, economic decline, and the social forces that accompany these events. In the years after the end of the generation-long Peloponnesian War in 404 BCE, the various Greek city states continued to war with one another in efforts to establish or resist hegemony, until Philip IV of Macedon finally took advantage of the weakened condition of the Greeks to conquer most of Greece by 328 BCE. Although his conquest wasn't particularly destructive as such things go, he did destroy entire cities, kill many thousands, and sap the strength of Greece. His son and heir, Alexander the Great, continued to weaken Greece by bleeding men and resources away in order to wage his wars of conquest. In the centuries that followed, Greece would rarely know peace, until the Romans finally conquered the Greeks in a bloody and destructive series of campaigns that ended in 86 BCE, the year in which the Roman general Sulla destroyed both the Academy and the Lyceum—as well as much of Athens. Although the Romans continued to travel to Greece because of the reputation of Greek teachers and the prestige Romans gained by knowing Greek and Greek philosophy, Greece increasingly became an impoverished backwater on the periphery of the Roman Empire.

Given these conditions, it's small wonder that Greek intellectual vigor declined—or that a much stronger trend in otherworldly, and sometimes world denying, philosophical and religious beliefs began to develop. First among these was Gnosticism, a collection of religions and philosophical movements that coalesced in the second century CE from influences that included Platonism and Pythagoreanism. Gnostics tended to be highly secretive about their beliefs and practices, and there was no single version of the movement. Thus, it's difficult to say anything definitive and comprehensive about Gnosticism, especially without going into far more depth than need detain us here. But in broad strokes Gnostics believe in a singular monadic deity (a monad is a philosophical term for an utterly unitary entity, such as the Prime Mover in Aristotelian philosophy), which emanates an illusory entity (though some Gnostics saw this entity as very real, and utterly opposed to the primary deity) known as a demiurge. This demiurge then creates an image of a higher-level reality, which is at best deeply flawed or at worst utterly evil, and is the sensible world around us. Because of this belief that the sensible world is either flawed or evil, Gnostics were world denying and often practiced moderate to severe ascetic (self-denial) practices in order to free the soul of bodily influences. A version of Christian Gnosticism developed, which among other beliefs held that the scriptures contain deep hidden messages only available to Gnostics, that Jesus was a spirit that only seemed to be human, and that all matter—including the body—is an evil to be escaped from. It should be no surprise that most Christians saw this form of Christianity as a heresy to be eradicated. Gnosticism appears to have risen after Christianity as a religio-philosophical movement and reaction against the problems the Greek world was experiencing.

Similarly, Plotinus (204–270) was an Egyptian Greek who settled in Rome where he taught for twenty years, during a time when the Roman Empire was itself in steep decline during a period known as the third Century Crisis. Between 235 and 284, the Empire experienced widespread plague and famine, as well as an economic crisis that almost resulted in financial collapse. These conditions made it difficult to resist barbarian incursions and enemies on her borders, as the army suffered numerous defeats and generals repeatedly rebelled and proclaimed themselves emperor, sometimes assassinating the sitting emperor— there were twenty-six emperors during this period, many lasting no more than six months. It was in this environment that Plotinus elaborated upon and developed Plato's thought into a system that was even more focused on otherworldly elements than his source material. Organizing his thoughts into a work known as the *Enneads*, Plotinus' philosophy posited that the sensible world is an emanation of the One, a completely unitary and perfect entity. This emanation occurs not as an intentionally creative act, but as something of an after-effect of the One's self-focused contemplation. The Intelligence then shapes this emanation into the sensible world, which is purely passive, through contemplation. The Forms that provide structure to matter are an image of the Forms contained within the mind of the Intelligence. Within this system, humans possess a higher and a lower-order soul, with the lower soul being the realm of desire, emotion, and such, while the higher soul yearns for reunion with the One from which it came, and can return through a process of intense contemplation. This system is remarkably otherworldly, but still within the tradition of Greek rationalism that privileges reasoning over observation. And its otherworldliness explains its attraction during such a troubled time. But the point is that many people were attracted to this system of thought because of the problems they experienced in the world, rather than this system being somehow imposed on them. And it would prove attractive for a long time; with the One baptized as God and the Intelligence as Jesus, Neoplatonism as it is known provided important philosophical underpinnings for Christianity beginning in a time known as the Patristic period, which we will explore in the next chapter.

In summary, the Greek discovery of reason and reason in nature occurred due to social and cultural forces present in many of the Greek city states that gave intellectual development a great deal of social prestige, which could be translated into political power in city states such as Athens. While it's true that all of these Greek city states were slave societies, there is no reason to believe that any given philosophical system was developed specifically to justify slavery, though Aristotle did provide a famous and influential justification for the system (basically he said that some people are born with souls suited for slavery, making it an obligation for born rulers to take charge of them). Additionally, since only the slaves and the poor did manual labor, the view of mechanical activity as a low-status pursuit would have acted to deter anyone from carrying out experiments, even if the idea had occurred to them. Furthermore, the idea that Platonism or Pythagoreanism suppressed a Greek focus on the material, sensible world is simply false—after all, Aristotle, with his intense focus on empirical observations came after Plato, not to mention the atomist Epicurus or the physician Galen. Instead, social forces that included the economic decline and conquest of Greece led to a shifting emphasis on belief systems that were otherworldly in their focus, but still often fell within the spectrum of Greek rationalism that privileged reasoning over observation.

Further Reading

Aristotle. Collected works, available at: http://classics.mit.edu/Browse/browse-Aristotle.html

Barnes, Jonathan. *Early Greek Philosophy*. New York: Penguin, 2001, 2nd edition.

Blackson, Thomas A. *Ancient Greek Philosophy: From the Presocratics to the Hellenistic Philosophers*. Malden: Wiley Blackwell, 2011.

Bruell, Christopher. *Aristotle as Teacher*. South Bend: St. Augustine Press, 2014.

Cartledge, Paul. *Democritus*. Abingdon: Routledge, 1999.

Clark, Stephen R. L. *Plotinus: Myth, Metaphor, and Philosophical Practice*. Chicago: University of Chicago Press, 2016.

Graham, Daniel. *Science before Socrates: Parmenides, Anaxagoras, and the New Astronomy*. Oxford: Oxford University Press, 2013.

Hendrix, Scott. E. "Natural Philosophy or Science in Premodern Epistemic Regimes? The Case of the Astrology of Albert the Great and Galileo Galilei." *Teorie vědy / Theory of Science: The Journal for Interdisciplinary Studies of Science*, 33.1 (2011): 111–132.

Kahn, Charles H. *Pythagoras and the Pythagoreans*. Indianapolis: Hackett Publishing Company, 2001.

Leroie, Armand Marie. *The Lagoon: How Aristotle Invented Science*. New York: Penguin, 2014.

O'Brien, Elmer. *The Essential Plotinus*. Indianapolis: Hackett Publishing Company, 1975.

Plato. Collected works, available at http://classics.mit.edu/Browse/browse-Plato.html

Sagan, Carl. *Cosmos*. New York: Random House, 1980.

Stanley, Thomas and James Wasserman. *Pythagoras: His Life and Teachings*. Lake Worth: Ibis Press, 2010.

Tuplin, Christopher and Tracey Elizabeth Rihll, eds. *Science and Mathematics in Ancient Greek Culture*. Oxford: Oxford University Press, 2002.

Tyson, Neil deGrasse. *Death by Black Hole: And Other Cosmic Quandaries*. New York: W.W Norton, and Company, 2007.

CHAPTER 2

Studying the Natural World in an Age of Faith

I n this chapter we'll explore the relationship between science—
or more accurately, natural philosophy—and Christianity in
the Middle Ages. This is an era in European history that Will
and Ariel Durant dubbed *The Age of Faith* in their encyclopedic
Story of Civilization, and with good reason. Catholicism reigned
supreme in Europe, providing the grounding for government, education, and society itself. With the exception of a small number
of Jews scattered throughout Europe and a smaller number of
Muslims on the Iberian Peninsula and in Sicily, to be a European
was to be a Catholic—or to be a heretic, and thus subject to prosecution and execution if found out. Because of the dominance of
Christianity during this period, the Middle Ages are something of
a ground zero in the conflict thesis. John W. Draper refers to the
medieval period as one of "semi-barbarian ignorance and superstition," when "the physical sciences were . . . discouraged or
perverted by the dominant orthodoxy" of Catholicism, in Andrew
Dickson White's words. In Carl Sagan's 1980 book, *Cosmos*, the
Middle Ages were simply a time when "the sciences of classical
antiquity had been silenced," which is why both versions of the
Cosmos television series ignore the medieval period, except for
a few remarks on the repressive nature of the Catholic Church
in this era. As we'll see, however, the evidence doesn't support

that view. Yes, there were extensive periods of limited intellectual output, but that was due to economic and political forces. Furthermore, while the development of science is decidedly slow during this period, that was again due to social forces that were sometimes identical to those found in ancient Greece. Those who took an interest in natural philosophy were most often teachers—science was not a profession and there was no particular prestige attached to the development of scientific ideas, and even less attached to applied science. Finally, what natural philosophy there was, was nurtured and supported by the church and by religious teachings.

In order to understand the relationship between religion and science in the Middle Ages, we have to begin our analysis with a few words about Christianity and Christian thinkers who came before. From the beginning, Christianity attracted followers from a range of social classes. Although at first it was largely a movement within the community of Hellenized Jews—meaning Jews who had adopted and been influenced by elements of Greek civilization and the Greek language spread to the Near East by Alexander the Great (356–323 BCE). The Apostle Paul (c. 5–c. 67), who was himself a Hellenized Jew, broadened Christianity's reach into the Gentile (non-Jewish) world, in spite of opposition from Peter the Apostle (died between 64 and 68). The Roman authorities largely ignored the new religious movement, until the Emperor Nero (37–68) blamed Christians for a fire that destroyed much of Rome in 64. Many Romans saw Christians as secretive and untrustworthy because of their habit of conducting ceremonies in private, and the Christian unwillingness to participate in the Roman state religion—most notably, performing acts of worship directed at the emperor—made them suspect in the eyes of Roman authorities. There followed over two centuries of periodic persecution of the Christians, with the worst period being from 303–306 during the reign of Diocletian (r. 284–305). When Emperor Constantine (r. 306–337) came to power, he restored legal rights to Christians and in 313 formally recognized Christianity as legal within the Empire.

Although Christians had always had an incentive to be educated enough to read their religious writings, it would not be until the second century that Christians began to develop an intellectual tradition of their own in response to arguments over doctrine within and attacks from those without the faith. The earliest of these Christians to do so lived in the Greek-speaking regions of the Empire in the East and were educated in the Greek philosophical traditions. The philosophies of Plato, and his later interpreter Plotinus (204–270), held the most attraction for the majority of these intellectuals, but as I noted in chapter 1 that was also true for non-Christian intellectuals during this period. Those who study this era of Christian history call it the Patristic period, because many of the Christian writers from these early centuries of Christianity are known as Church Fathers (*pater* is the Latin word for "father," which is where the term "patristic" is derived from) because of their tremendous influence on the development of Christianity. The earliest of these writers were men like Justin Martyr (martyred in Rome sometime between 162 and 168) and Clement (c. 150–c. 215), a teacher in Alexandria. Both of these men argued that anyone could develop ideas that were true, because all people had been created by God, and therefore all people—even pagans—are capable of producing ideas consistent with divine truth. Furthermore, scholars like Justin and Clement argued that Greek philosophy was necessary in order to allow people to steer clear of bad or heretical ideas, and they were both pragmatic enough to recognize that if Christians weren't well educated, Christianity would be labeled backwards and without intellectual grounding.

This isn't the place to analyze the Patristic tradition in depth, but since these writers would have such an outsized influence on Medieval Christianity, we need to understand what some of these men had to say in order to clarify the relationship between science and religion. One such writer we need to consider is a Carthaginian (in northern Africa) scholar, Tertullian (c. 155–230), because he is often held up as an exemplar of an anti-intellectual Christian scholar. In his *Prescription Against Heretics* written in the early third century, Tertullian famously stated that,

> when the apostle [Paul] would restrain us, he expressly
> names *philosophy* as that which he would have us be on
> our guard against. Writing to the Colossians, he says,
> "See that no one beguile you through philosophy and vain
> deceit, after the tradition of men, and contrary to the wis-
> dom of the Holy Ghost." He had been at Athens, and had
> in his interviews (with its philosophers) become acquainted
> with that human wisdom which pretends to know the
> truth, whilst it only corrupts it, and is itself divided into
> its own manifold heresies, by the variety of its mutually
> repugnant sects. What indeed has Athens to do with
> Jerusalem? What concord is there between the Academy
> and the Church? what between heretics and Christians?
> Our instruction comes from "the porch of Solomon," who
> had himself taught that "the Lord should be sought in
> simplicity of heart." Away with all attempts to produce a
> mottled Christianity of Stoic, Platonic, and dialectic com-
> position! We want no curious disputation after possessing
> Christ Jesus, no inquisition after enjoying the gospel! With
> our faith, we desire no further belief.

The statement often paraphrased as "what has Athens to do with Jerusalem?" refers to the distance between Greek philosophy and Christian faith, and is often used to indicate that Tertullian—and by extension, Christianity in general—was uncompromisingly opposed to Greek intellectual traditions.

In the same work, Tertullian also turned his rhetorical gifts directly toward an attack on natural philosophers, when he asked,

> Now, pray tell me, what wisdom is there in this hankering
> after conjectural speculations? What proof is afforded to
> us, notwithstanding the strong confidence of its assertions,
> by the useless affectation of a scrupulous curiosity, which
> is tricked out with an artful show of language? It therefore
> served Thales of Miletus quite right, when, star-gazing as
> he walked with all the eyes he had, he had the mortification

of falling into a well. His fall, therefore, is a figurative
picture of the philosophers; of those, I mean, who persist in
applying their studies to a vain purpose, since they indulge
a stupid curiosity on natural objects, which they ought
rather [intelligently to direct] to their Creator and Governor.

The phrase, "useless affectation of a scrupulous curiosity" of those "who persist in applying their studies to a vain purpose" seems to be a clear enough statement of opposition to Greek philosophy—but is it?

Tertullian clearly thought that the study of the Scriptures was more important than the study of philosophy. However, in this work he is specifically attacking heresy—especially Gnosticism—which he saw as sometimes being born out of philosophy. When not directly engaged in attacking heresy, Tertullian could sound far different when writing about education in general, or philosophy in particular. In his work *On Repentance*, Tertullian echoed Justin Martyr and Clement by stating that "reason . . . is a thing of God, inasmuch as there is nothing which God the Maker of all has not provided, disposed, ordained by reason-nothing which He has not willed should be handled and understood by reason." He even went so far as to argue that the study of nature could lead to the knowledge of divine truths, such as the immortality of the soul. Furthermore, Tertullian regularly applied the tools of Aristotelian logic in his writing, demonstrating that he had a pragmatic approach to the value of Greek philosophy.

The most influential of the Church fathers, though, was another North African, Augustine (354–430). Born to a Christian mother and a pagan father in Thagaste, Augustine received a thorough education at the North African schools of Thagaste, Madaura, and Carthage before going on to teach rhetoric in Carthage, Rome, and Milan (which was the capital of the western half of the Roman Empire at this time). In his youth, Augustine demonstrated the validity of Tertullian's fears by viewing Christianity as lacking in intellectual foundations, and instead found himself drawn to the dualistic religion Manichaeism. Eventually, he found this faith wanting and spent some time focused on Neoplatonism. However, while teaching in Milan he became acquainted with the highly educated Bishop of Milan, Ambrose (c. 340–397). Augustine found his Neoplatonic form of Christianity deeply attractive, and in time converted and eventually became the bishop of the North African town of Hippo.

Far from desiring to substitute reason with faith, Augustine wrote to the grammarian Consentius (fifth century) a rather strong statement in support of reason: "Heaven forbid that God should hate in us that by which he made us superior to the other animals! Heaven forbid that we should believe in such a way as not to accept or seek reasons, since we could not even believe if we did not possess rational souls." He did note that "faith must precede reason and purify the heart and make it fit to receive and endure the great light of reason," and vowed in his work, *Against the Academics,* "never to deviate in the least from the authority of Christ." However, this in no way meant that Augustine did not value the classical intellectual tradition. Well before he was a bishop—or a Christian—Augustine was a teacher of rhetoric, trained in the best Roman traditions. Although John W. Draper states that Augustine's "works are an incoherent dream," nothing could be further from the truth. They are structured according to the rules of logic and he often anticipates modern philosophical positions by many centuries. For example, he discusses the basis for knowledge in his autobiography, *The Confessions,* where he chooses a beginning point in the simple fact that "if I am mistaken, I am," which is the same starting point Descartes (1596–1650) chooses more than a millennium later, when he writes, "I think, therefore I am." As the French medievalist Etiene Gilson put in his study on *The Christian Philosophy of Saint Augustine,* "one never knows if Augustine speaks as a philosopher or a theologian," because the two pursuits are inextricably interwoven in Augustine's works.

But what about Augustine's position regarding natural philosophy? Did his "doctrines . . . have . . . the effect of . . . placing theology in antagonism with science," as Draper said, leading to "theological views of science which have never led to a single truth," as Andrew Dickson White, put it? Augustine certainly didn't evidence a particular interest in the natural world, but neither did he suggest it should be ignored. In his book *On Christian Teaching,* Augustine does definitively state that one should only study those things that will help humans understand God better. But what fields contribute to a better understanding of God? Augustine lists mathematics and geometry, rhetoric and logic, astronomy, physics, and even music—in short, he enumerates every area of what were the classical liberal arts. He argues that humans can't know God directly, so how can people come to know him? By studying his creation—in other words, everything around us as well as all the ideas and abstract pursuits of which humans are capable. After all, in Augustine's view all things can ultimately be traced back to God. And this is exactly how he approaches his own work. In his *Literal Commentary on Genesis,* Augustine seeks to parse out how the creation of life as laid out in Genesis progressed. Far from taking what a modern fundamentalist would see as a "literal" approach to the text, Augustine argues on theological grounds that a six-day act of creation was inconsistent with other passages in Genesis. Instead, he went on, this framework was a logical rather than a physical one. He elaborates that God infused the Earth with seminal ideas—the basic Forms of life—which developed first into the simplest forms of life, then later into increasingly complex forms. Students reading Augustine's account are often shocked with how much it sounds like Darwinian evolution, a point that hasn't been lost on the Catholic Church; many Catholic theologians in the modern period—such as Pope John Paul II (1920–2005)—have referred to Augustine's analysis when expressing their own support for evolutionary theory.

It wasn't only his understanding of Genesis that proved to be influential, though. In fact, one reason it's worth spending a bit of effort in coming to understand Augustine in any discussion about the Middle Ages, is because his model for understanding the bible influenced the Catholic approach to scripture down to the present day. In *On Christian Teaching* Augustine draws on Galatians 4:21-26, where Paul interprets a story from the Old Testament as an allegory to support his argument that the bible must often be interpreted allegorically. The historical meaning, as Augustine would call it, is but one way of understanding any given biblical text, and if a reader stops there, then he or she is missing the true meaning of the text. We should keep this point in mind as we think about religion and its relationship to natural philosophy or the development of science, particularly in the Middle Ages; every monastery would have had copies of Augustine's works, and every theologian and religious official from the parish priest to the pope would have read Augustine and learned to understand the bible according to the model Augustine proposed.

In some ways Augustine died along with the Roman Empire in the West. In 430, a group of barbarians whose names have become synonymous with destruction, the Vandals, were besieging the city of Hippo while its bishop—Augustine—lay dying within its walls. It's probably fortunate for him that he died before they could take the city. Only forty-six years later, in 476, another barbarian forced the last Roman Emperor of the west, an entirely ineffectual teenager named Romulus Augustulus (460–527), to retire. Although Roman institutions such as schools would survive for a time in Italy and other parts of what had been the Roman Empire in the West, they either slowly deteriorated or were destroyed during cataclysmic events such as Emperor Justinian I's (483–565) conquest of the west. As the emperor of what he thought of as simply the Roman Empire— though he truly only controlled the eastern half of the lands of the empire at its peak, the territory most commonly known as the Byzantine Empire—Justinian saw it as his duty to retake the lands lost to various barbarian groups in the west. Therefore, he sent his most talented general, Belisarius (500–565) to do so. The Gothic Wars (535–554) that followed briefly brought much of what had been the western Roman Empire under Justinian's control, though his control of a reunited empire would be brief. However, the Gothic Wars also devastated much of Western Europe, destroying the vestiges of Roman institutions and infrastructure that had survived so effectively that literacy itself was almost lost to Europe.

Medieval historians generally avoid the term "Dark Ages," which is sometimes applied indiscriminately to the entire period from the end of the Roman Empire in the west until the Renaissance began at the end of the fourteenth century. But if the term is to be used, it should be reserved for the centuries immediately following the collapse of Roman power in the west. There were occasional bright spots to be found on the continent— Visigothic Spain maintained Roman institutions into at least the seventh century, and while Anglo-Saxon England lacked Roman influences almost entirely, flourishing vernacular literature and scholarship flourished on the island until the time of the Norman Conquest in 1066. However, for the most part things could hardly have been worse. Petty kings waged a seemingly endless series of wars against one another, the vast majority of people were too focused on survival to worry themselves with anything so lofty as education, and there was hardly any surplus food to sustain teachers or those who would learn from them. Only in the monasteries scattered across Europe was the flickering light of learning kept alive. In these places, monks following the Rule of St. Benedict (480–547) grew their own crops, tended the flocks of sheep that provided the vellum with which one could create books, taught one another, made their own books, and in other ways practiced self-sufficiency. But even there the flame of learning burned perilously low; libraries rarely had more than a dozen or two books and the monks focused almost entirely on copying and preserving the handful of works they had, rather than creating anything new.

Very little of the Roman legacy of scholarship survived in Europe, and almost nothing of the vast output of the Greeks and their Hellenistic was available. What little was available was only found in paraphrases or snippets taken from larger works and translated into Latin, for it was almost impossible to find anyone in Europe who could read Greek during the early Middle Ages. Rudimentary elements of Aristotle's logic could be found in translations done by the fifth century Italian writer, Boethius, and the broad outlines of Plato's thought could be discerned by reading Augustine. As for natural philosophy, little was available beyond the odds and ends the first century Roman writer Pliny had worked into his encyclopedia *Natural History*. It would be easy to blame the monastic devotion to God for this loss of Greek natural philosophy, but really it was the natural inclination to focus on not starving or being killed by the marauders from next door that kept most Europeans too preoccupied to worry about anything like science. Rather than blame the monks for not doing more, we should thank them for preserving what they did.

However, not all of Europe was as poor and fraught with danger as the European heartland. In the seventh century the prophet Muhammad had begun to preach a message of universal brotherhood among those who recognized that "there is no god but God, and Muhammad is his prophet," transforming the warring tribes of the Arabian Peninsula into an organized and expansionistic unified force under the banner of Islam. As this *umma*—brotherhood—of Muslims spread outwards, they soon conquered the Sasasanid Persian Empire, which was centered on modern-day Iran and had been weakened by centuries of warfare with the Byzantine Empire. The Byzantines—which represented the eastern half of what had once been the Roman Empire (and they still referred to themselves as simply *Romanoi*, meaning "Romans")—had been equally weakened, and parts of their territory was soon consumed by the spreading forces of Islam.

The spread of Islam is a fascinating story in its own right, but for our purposes the most important thing is that as territory fell to the Muslims, they soon controlled vast swaths of territory peopled by members of ancient and highly literate cultures. In the former Byzantine lands of what today is Syria and Palestine, this included Nestorian Christians who were overjoyed to be a protected minority under the Muslims, instead of persecuted as heretics by the Byzantines for their rejection of the Trinity. These Nestorians were fluent in the Greek that was still the daily language of the Byzantine Empire, and included many highly educated individuals who soon found employment translating Greek philosophical works into Arabic in places like the various Houses of Wisdom, which combined a school of translation with a school of advanced study, the most famous of which was located in Baghdad. The Muslims tended to be interested in elements of Greek thought that they saw as useful, so their interest in poetry and theater was limited, but they were extremely interested in natural philosophy.

This interest was fortunate, for it insured the survival of a great deal of Greek writing on natural philosophy as well as its spread. From the standpoint of a modern reader, it may seem curious that one of the founders of the conflict thesis, John W. Draper, has much good to say about Islam and its spread, for today unfortunate and uninformed views about Islam and its history are all too common. Nevertheless, Draper was very aware of the positive contributions of the Islamic people to the preservation and spread of Greek philosophy, further reinforcing that it wasn't really "religion" that roused his anger, so much as Christianity. In his *History of the Warfare between Religion and Science,* he wrote that the Muslim lands were

> dotted all over with colleges . . . in Mongolia, Tartary,
> Persia, Mesopotamia, Syria, Egypt, North Africa, Morocco,
> Fez, Spain. At one extremity of this vast region, which far
> exceeded the Roman Empire in geographical extent, were
> the college and astronomical observatory of Samarcand, at
> the other the Giralda in Spain. . . . The superintendence
> of these schools was committed with noble liberality
> sometimes to Nestorians, sometimes to Jews. It mattered
> not in what country a man was born, nor what were his
> religious opinions; his attainment in learning was the only
> thing to be considered. The great Khalif Al-Mamun [who
> ruled the Abbasid Empire, which stretched from modern-
> day Afghanistan into Spain at its peak, in the early 9th
> century] had declared that "they are the elect of God, his
> best and most useful servants, whose lives are devoted to
> the improvement of their rational faculties; that the teach-
> ers of wisdom are the true luminaries and legislators of
> this world, which, without their aid, would again sink into
> ignorance and barbarism."

His writing might be a little florid and his diction a little dated, but in this instance, Draper was essentially correct. Muslim rulers took Muhammad's command to "seek knowledge unto China" to heart, and were liberal benefactors of scholars and religious institutions.

Well informed Europeans weren't unaware of the higher level of learning and Greek philosophy to be found in the Muslim lands that were far wealthier and more secure than their European neighbors. Italian merchants traded with Muslims and were certainly aware that mathematics was far more advanced in Islamic territories than in Latin Christendom, the European lands where Catholicism reigned supreme. However, the first record of a scholar gaining access to the wealth of Islamic natural philosophy was a tenth century Frenchman named Gerbert of Aurillac (c. 946–1003). As a young monk in southern France who had a talent for mathematics, his abbot arranged for him to travel into Catalonia in the party of Borrell II, the Count of Barcelona. This region of Catalonia in northeastern Spain had been retaken from the Muslims who had conquered most of the Iberian Peninsula in 711. But unlike most of Europe, Muslims were tolerated there due to their usefulness as much as their numbers, and the monastery of Vic—just to the north of Barcelona—where Gerbert stayed, received manuscripts from monks who lived in Islamic controlled regions of modern-day Spain. It was through his writings that the abacus and the armillary sphere (an astronomical tool) were reintroduced to Europe, and Gerbert also wrote on the decimal numeral system using Arabic numerals, rather than the cumbersome Roman numbers used in the rest of Europe. Gerbert would later become Pope Sylvester II in 999, and would use his office to promote the study of mathematics and the natural philosophy available in Muslim-controlled Spain.

By the time Sylvester II took the papal seat, things were improving in Europe. By the ninth century rulers had begun to emerge who could bring a measure of peace to larger numbers of people, with Charlemagne (ruled from 768–814) being the most famous. In the tenth century agricultural improvements increased the amount of available food, creating a surplus that allowed the wheels of commerce and banking to grind slowly into gear by the eleventh century as those who controlled the land sold or bartered their surpluses. The result was a growing demand for products produced in the newly developing cities. Thus, by the end of the eleventh century there was enough peace and prosperity for a sliver of the population to begin to think about something other than the basics of survival. A small number of these people lived in cities where they manufactured goods, bought and sold them, or practiced other urban pursuits, such as education. While these developments were important for many reasons, one important factor concerned the church. By the end of the tenth century, it was clear that the Catholic Church was in dire need of reform, and many spoke out about the issues caused by illiterate priests at the lowest level to popes with little or no religious sensibilities at the highest. The story about why this reform was so desperately needed is a fascinating one, involving figures such as the teenaged Pope John XII (r. 955–964) who engaged in various sexual excesses and may have been murdered by a jealous husband. But what is most pertinent to our current study is that as the numbers of educated people grew during the high point of the Middle Ages, these educated people continued to be almost universally in orders, either secular clergy—those who are expected to operate in the world—monks, or one of the new orders of friars that formed in the thirteenth century.

The association between the church and education in the Middle Ages is well known to those with an interest in the period, and has been used as a point of attack by many. John W. Draper sums the entire Middle Ages up as a period when "no one could indulge in freedom of thought without expecting punishment." For him, as well as his contemporary Andrew Dickson, the medieval church controlled education as a means to control thought and repress science. More recent writers have agreed, with Carl Sagan's 1995 work, *The Demon Haunted World,* characterizing the Middle Ages as a period when the church kept Europeans focused on issues relating to religious minutiae or superstitious musings about demons rather than developing scientific understandings of the world. For these writers, the fact that almost all educated people in the Middle Ages were in orders was a point against, rather than for, the church. However, there is another side to this story.

As available food resources expanded and European nations grew in power enough to create increasing security for the people of Europe, there were more resources available for the support of specialists of all kinds, including teachers and students at the schools growing up around cathedrals that were the precursors to the universities that would emerge at the end of the twelfth century. These schools were aimed at educating the clergy, as Christian intellectuals from the time of Augustine forward had warned of problems that could emerge within the church as the result of an ill-educated clergy. These problems were apparent not only at the level of the papacy, which would largely correct the problems of the tenth century in the eleventh, but also at the level of the local parish priest. Many of these men knew no Latin and were forced to memorize the phrases necessary for the rites of the Catholic Church. Even worse, they lacked the knowledge to differentiate between theologically correct notions and heresy, and they certainly couldn't fully understand the wonder of God and his creation, the universe.

In order to correct this situation, it wasn't enough to establish schools. These schools required books and the knowledge they contained. Therefore, scholar-adventurers such as the Italian monk Gerard of Cremona (c. 1114–1187) traveled into Muslim-controlled Spain in order to acquire the knowledge of the Greeks and those who wrote about Greek knowledge in Arabic. A few, such as Gerard, may have learned Arabic. However, most worked with local scholars in places like Toledo in order to render these works into Latin. Many of these Spanish scholars were Jews, who were required by the tenets of their faith to be knowledgeable in Hebrew, but used medieval Spanish and Arabic for their day-to-day activities, making them comfortably multi-lingual. Therefore, translators frequently rendered works written by authors, such as Aristotle in Spanish, before translating them into Latin, and sometimes these books would first be translated into Hebrew before being translated into Latin. Thus, any given work might go through four levels of translation before becoming comprehensible to the educated elite of Europe, for whom Latin was the language of learning.

This process led inevitably to problems. For example, Abū Alī al-usayn ibn Abd Allāh ibn Al-Hasan ibn Ali ibn Sīnā—who thankfully is better known in the West as Avicenna, a Latinized form of "ibn Sina," which simply means "son of Sina"—was a Persian scholar of staggering intellectual capabilities who wrote more than 450 books, almost entirely on scientific topics such as astronomy and mathematics in the tenth century. He was most interested in medicine, though, and it would be his *Canon of Medicine*, a five-book encyclopedia of medicine blending Greek, Hellenist, and Arabic ideas, that would make him most famous. Translated into Latin in the thirteenth century, this work became the standard textbook for medical students for the next 400 years. It contains much that is useful, such as a detailed consideration of smallpox that is remarkable for its commitment to empirical observation, but the process of its translation into Latin was not always smooth. There was no such thing as an Arabic to Latin dictionary, and while Latin is a very compact language, Arabic is not. It takes many more words to say something in Arabic than in Latin. Contained within a discussion of stomach ailments, medical students would learn that the preferred treatment was to dose the patient with poisonous substances. There is a certain sense to this, as a deadly poison in small doses has a purgative effect, and this purgative effect can make someone with an upset stomach feel better. This was especially true in Europe where, as a sign of status that would persist well into the eighteenth century, those who could afford to do so would ordinarily eat meals that consisted of little more than several pounds of meat. It's probably no surprise that ferociously uncomfortable constipation was common among wealthier Europeans, and a touch of poison could be remarkably effective at loosening up bowels clogged by too much fatty meat and too little fiber. However, a hair too much of the treatment could cause death, which many saw as an unfortunate side effect. It wasn't until the Renaissance when a scholar as knowledgeable in Arabic as he was in Latin and Greek noticed that the Arabic word that had been mistranslated as various poisonous compounds actually meant "peppermint," an oil which has a soothing effect on upset stomachs without the danger of killing the patient. However, this correction came too late to change medical practices, and physicians continued to use poisons such as arsenic for a range of disorders from stomach ailments to dental cavities into the eighteenth century.

In spite of hiccups such as that one, the translation of Greek and Arabic works into Latin during the twelfth century contributed immeasurably to European intellectual life, and that included the study of natural philosophy. These translations occurred during a period known as the twelfth Century Renaissance, and one component of that explosion of intellectual activity was what scholars such as Steven A. Epstein have termed the "medieval discovery of nature." What this term refers to is that men such as the twelfth-century French monk, William of Conches wrote about natural phenomenon from pregnancy to psychology without appealing to God or his divine power to describe these phenomena or their possible causes. Instead, he used Aristotelian philosophy and the ideas of Constantine the African, a North African physician who had moved to Salerno, Italy, and converted from Islam to Christianity in the eleventh century. William offered explanations for lightning and the way in which air becomes both thinner and colder as one climbs higher up a mountain that sound strikingly modern, based an evaluation of cause and effect that is grounded in physical processes. And William was far from an outlier, as monks who applied philosophy in an effort to understand the natural world abounded in the twelfth century. The English monk, John of Salisbury, took this approach in his analysis of politics when he made the idea that the human capacities of reason and language are what transformed humans from primordial animalistic beings into beings capable of freely willed political associations a central element of his works on politics and education, while Honorius of Autun's encyclopedic *Image of the World* passed along the basic Greek and Roman geographical and cosmological theories, complete with naturalistic discussions of cause and effect in the world, in such a useful format that it would become a staple of medieval libraries. True, Honorius also discussed such things as guardian angels, but when it came to a consideration of the physical world he stuck to naturalistic reasoning.

There were certainly ways for a twelfth-century scholar to get into trouble by engaging in academic discussion. Peter Abelard was the most famous teacher of the day, with a reputation enhanced by his massive ego, a complete absence of hesitation to self-publicize, and a notorious love affair with a beautiful young Parisian woman named Heloise that resulted in a pregnancy, secret marriage, and ultimately Abelard's castration at the hands of thugs in the employ of Heliose's uncle. After his castration and subsequent decision to become a monk, Abelard decided to apply Aristotelian logic to an analysis and explanation of the Trinity. Due as much to the fact that few could understand what on earth Abelard was going on about as any truly heretical content, his book *On the Trinity* was condemned as heretical and Abelard was ordered to condemn the book to the flames with his own hands and to desist from writing about or teaching on the Trinity. Convinced that he simply hadn't explained himself well enough— and like philosophers from his day to the present, it really was pretty difficult to understand his writings—Abelard made another pass at explaining the Trinity late in his life. This second version of *On the Trinity* was no better received than the first, and Abelard died while under genial and comfortable house arrest.

However, we should note that Abelard got himself into trouble for attempting a logical explanation of the Trinity, not for writing natural philosophy. And while it was theoretically possible that one could get into hot water by writing natural philosophy—it wasn't a good idea to call the virgin birth of Christ into question, and one had to be careful when writing about the Eucharist, for example—there really aren't any examples of scholars getting into trouble for their writings on natural philosophy. In fact, such writings were something of a fad in the twelfth century, and everyone from the pope to local parish priests could be expected to have read works dealing with phenomena like abnormal births, lightning, meteors, or rain, to name only a few examples that were popular topics of the day. And don't forget that not only were monks and priests *reading* these works, but it was also monks and priests who were *writing* them. Few who were not in orders were literate in the twelfth century. So rather than religion—either institutional or as a private faith—repressing science in the twelfth century, it was instead a driving force. That raises the question of why, though, and the answer is deceptively simple. On the one hand, theologians understood that there was no way to study God other than through study of his creation, the universe, and all that it contained. On the other hand, there was a great interest in miracles, but concern lest an event be mistakenly labeled a miracle. How else could one know the difference between a miracle and a naturally occurring event if not by understanding how things work in the physical world? Everything looks like a miracle to those ignorant of the natural processes of cause and effect, after all.

The point is that in the twelfth century the Catholic Church was the primary provider of institutional support for scholars, and Catholic theology motivated these scholars to analyze the natural world. Rather than suppressing science, the church and its theology actively promoted it. But what about the following centuries? After all, Pope Gregory IX (r. 1227–1241) established the papal inquisition in 1231 and the pursuit of heretics took off in earnest. Before this date the pursuit of heretics was left up to local bishops who often held their positions in absentia, were cozy with local political elites who frequently had little interest in seeing heretics pursued, and were busy with a job that combined spiritual responsibilities with the sort of mundane administrative issues that modern bishops rarely have to consider, such as the gathering of grain and the distribution of bread to the poor. Thus, while there is no doubt that the thirteenth century saw some very prominent heresies—such as the Albigensians who held a dualist faith in which Satan and God were equally powerful—the uptick in the prosecution of heretics likely had more to do with an improved policing apparatus than a growth in the absolute numbers of heretics. Furthermore, the church itself was becoming increasingly centralized, self-confident, and powerful, so if there was a suppression of science by the medieval church, surely it came out of this period.

Andrew Dickson White certainly thought that the medieval inquisition, as he called it—by which he meant the papal inquisition—was responsible for the repression of science and scientists. In his heroic account,

> men like . . . Roger Bacon . . . cultivated sciences . . . in
> spite of charges of sorcery, with possibilities of imprison-
> ment and death, [who] kept the torch of knowledge burning,
> and passed it on to future generations.

What a hero, who took a stand against the oppressive might of the medieval church! Or is it quite that simple?

So, who was Roger Bacon (c. 1220–c. 1292)? Born in Ilchester in Somerset, England, Bacon studied at the University of Oxford, which was in its infancy at the time. Universities are one of the medieval church's gifts to the modern world, growing out of the cathedral schools that had become prominent in the twelfth century. The oldest university is either the one at Paris or Bologna, both of which date to the late twelfth century. Lacking any physical infrastructure at first, universities were church-run guilds of scholars who came together to offer a more well-rounded education than that found at earlier institutions. Where once students would have studied primarily under one master and learned more or less whatever he felt like teaching, universities offered a standard curriculum based largely on the works of Aristotle, at least until one reached the level of doctoral studies. Students entered the university around the age of fourteen or fifteen, having already mastered Latin because all instruction was in that language. The curriculum for the Bachelor of Arts degree required a student to demonstrate familiarity with the seven liberal arts, grammar (the Latin language, that is), rhetoric, logic, music (theory, not application), mathematics, geometry, and astronomy. In 1215 the fourth Lateran Council of the Catholic Church declared that all priests must have a bachelor's degree from a university (though in practice, this was not always true), so the B.A. definitely opened up career options. It's also worth noting in any consideration of religion and its relationship to science that the medieval church thought priests should, in effect, have the equivalent of a degree in math and physics. Once a student had earned the B.A., he (and it was always a "he," for these young men were considered priests-in-training) could go on for the Master of Arts, which would require a mastery of these subjects. If he went this far, he could then teach at a university or go on for one of the doctorate degrees, in medicine, law (either civil or church canon law, or both if he felt like showing off), or the field of study considered both most difficult and most prestigious, theology. Although Roger Bacon would come to be known as *Doctor mirablis* after his death, he only went so far as to earn a Master of Arts degree.

Bacon would have fit into university life at Oxford well. There were scholarships to support poor but intelligent young men. As only one example, in the centuries to come Tommaso Parentucelli was a boy raised only by his mother, after his surgeon (not a very high-status job at the time) father died, leaving too many children to care for and too little money. Yet thanks to such a scholarship, he earned his doctorate in theology before going on to become Pope Nicholas V from 1447 until his death in 1455. But Bacon, like most university students, was from a family of means, which likely meant from the lower nobility, though there were merchants in the thirteenth century who did quite well for themselves. The point is that students who attended universities were almost always from the sorts of families for whom manual labor was anathema. Furthermore, university students such as Bacon partook in a curriculum that built almost entirely around Aristotle's works, and Aristotle—like all free Greeks of his day—had a very low opinion of manual labor. In book one of his *Politics* he wrote that every professional, from a flute player to an artisan or laborer, was in "partial or limited slavery;" while he was not technically a slave, he lacks the leisure time to practice virtuous actions, and thereby to attain true virtue. Aristotle's master, Plato, was if anything more negative in his opinion of manual labor. In his dialogue on *The Laws* and *The Republic* he describes men who make money by any means other than inherited land as "low caste" and "cringing," as they exist in a depraved state. Bacon and his fellow students who came from wealthy backgrounds— Albert the Great was the son of the Count of Bollstadt in modern-day Germany, and his most famed student, Thomas Aquinas

(1225–1274) was the son of the Count of Aquino in modern-day Italy—must have been quite sniffy as they complimented one another upon pursuing the life of the mind as university masters, and pleased to find that Aristotle and Plato reinforced their bias against manual labor. Any impoverished student who managed to slip in the doors through the support of a scholarship would have rapidly adopted such attitudes lest he be ostracized by his higher-status colleagues. Needless to say, intellectuals within this social environment were unlikely to think of experiments as a worthwhile pursuit.

As a university master, Bacon lectured on a range of subjects, from Latin grammar to mathematics and astronomy at both the University of Oxford and the intellectual center of Europe in the Middle Ages, the University of Paris. While doing so, he developed quite a reputation for his caustic personality and regularly attacked even the most respected intellectuals of his day. For example, Albert the Great (c. 1200–1280)—who was known as "the Great" even in his own lifetime, for his command of Aristotelian philosophy—is now the Saint of Scientists for the Catholic Church and historians esteem him one of the greatest minds of his day. Albert was a Doctor of Theology at the University of Paris while Bacon was lecturing as a master there. Yet Bacon referred to him as "useless as a teacher" who was "ignorant" of the most important elements of natural philosophy. It's interesting that Bacon would level such an attack, for while Bacon struggled to write his *Opus maius* (*Great Work*), which he promised would weave natural philosophy and theology together, Albert wrote literally dozens of works on everything from astronomy to theology.

Bacon's attacks on Albert may have had as much to do with professional rivalry as anything else; Bacon joined the Order of Friars Minor (most commonly known as the Franciscans) in his 40s, while Albert was a member of the Order of Preachers (most commonly known as the Dominicans). Yet we should remember that Bacon never missed an opportunity to ridicule his opponents, for his fate likely had as much to do with the many personal animosities he inspired as anything else. It's also possible that he was simply jealous of the respect that scholars such as Albert received, for while Bacon produced a number of short works—most notable a work on *Perspective* (what today we would call optics) and his study *On Experimental Science*—these were actually short pieces of the larger work he attempted unsuccessfully to complete.

However, Bacon's reputation among modern scientists is built on these short essays, for which he is sometimes hailed as, to use the title of a book written by H. Stanley Redgrove in 1920, *The Father of Experimental Science*. This reputation is built on statements Bacon made such as his essay *On Experimental Science*, part of his never-completed *Opus maius,* that

> there are two modes of acquiring knowledge, namely, by reasoning and experience. Reasoning draws a conclusion and makes us grant the conclusion, but does not make the conclusion certain, nor does it remove doubt so that the mind may rest on the intuition of truth, unless the mind discovers it by the path of experience.

I should note that while the 1928 translator of the above selection rightly translates the Latin *experimentia* as "experience," that isn't always the case. Many look at that word and think "experiment," which, frankly, is rather wide of the mark. While medieval scholars interested in natural philosophy sometimes went to the effort of looking at the world around them, such activities were nothing like a modern scientist's experiments. For example, Bacon's contemporary, Albert the Great, tried unsuccessfully to feed stones to chickens in order to test a statement Aristotle had made about birds eating stones, and spoke to miners about stones when writing about them, but that's as close as you're going to find to an experiment performed by a medieval scholar. Medieval scholarship was an almost entirely bookish pursuit, wherein to prove something meant to construct a proper logical argument about the point under consideration. Logic, not experimentation, was the gold standard of proof in the Middle Ages.

That point should be kept in mind when we encounter statements such as Bacon's that

there are four great sciences, without which the other
sciences cannot be known nor a knowledge of things
secured . . . Of these sciences the gate and key is mathematics . . . He who is ignorant of this [mathematics] cannot
know the other sciences nor the affairs of this world.

This is precisely the kind of statement that has led writers such as Brian Clegg—who has an undergraduate degree in
Natural Sciences, with an emphasis on experimental physics,
from Cambridge and a master's degree in the mathematical
discipline of Operational Research from Lancaster University—
to call Bacon *The First Scientist,* which was the title of Clegg's
2001 biography of Bacon. However, there are real problems with
that view of Bacon. First, and most importantly, it muddies the
water when thinking of Bacon, for it causes us, consciously or
unconsciously, to assume that Bacon worked as a *scientist,* and
since we think we know how a scientist goes about his or her
job by modern standards, we erroneously assume that historical
personalities acted the same way. Which leads us to the second
point—while a modern person trained in the sciences, such as
Clegg, might assume that Bacon means what we do by terms
like *experimentia*—often translated as "experiment"—or when he
writes of mathematics, nothing could be further than the truth.
When Bacon wrote about experience, he was acknowledging that
ideas about the natural world should match up to our experience
of the natural world, but such experience included the obvious-
to-him fact that heavier objects fell faster than lighter ones, in line
with Aristotelian physics. And as for Bacon's view of mathematics, he saw this as a metaphysical discipline that dealt primarily
with higher order—that is non-material—existence. He certainly
didn't mingle mathematics with analysis of the natural world, so
he meant something very different by his statements about mathematics than a modern person unaware of the historical context
would assume that he meant.

So if Bacon wasn't a scientist in the modern sense of the word,
what about the claim that he was persecuted by the church for

his study of the natural world? Traditionally, it is held that Bacon died while imprisoned for "suspected novelties" around the year 1292. However, as Thomas Maloney, the editor and translator of Bacon's *Compendium of the Study of Theology* points out, the earliest reference to Bacon being imprisoned occurs in 1370—some eighty years after his death. While it's clear that something happened to Bacon late in his life, as he quit writing after the 1270s, it's not at all clear what that was or why it occurred. He may have been placed under a form of house arrest, but the most likely reasons for this are Bacon's interest in prophecies about the end of the world, his taking the side of the "spiritual Franciscans" (those in the Franciscan order who maintained a complete commitment against owning property, a position that was deemed heretical in the last quarter of the thirteenth century), a refusal to gain permission from his superiors in the Franciscan order before publishing, and the fact that he managed to make enemies of most European scholars and members of the hierarchy of the church. A man with who promoted alarmist stories about the end of the world with many well-positioned enemies could easily find himself in trouble with the authorities without it having anything to do with his interest in natural philosophy. There is no mention anywhere of opposition on the part of the papal inquisition, the papacy, or the church in general to any of his so-called scientific writings.

There were moments when members of the hierarchy of the Catholic Church would attempt to put a stop to avenues of scholarly speculation it found uncomfortable. The most famous such instance occurred in 1277, when the Bishop of Paris, Etienne Tempier (1210–1279) issued a list of 219 propositions forbidden for anyone at the University of Paris to teach or even think about. That sounds pretty extreme and seems to be definitive proof of the oppressive nature of the medieval church . . . or at least it does at first glance. The propositions were thrown together rather hastily by a committee that had taken less than two months to fulfill Pope John XXI's (1276–1277) mandate that Tempier investigate rumors of heresy at university. The result is an absence of organizational structure, imprecise language, and frequent contradictions, such as condemnation 93 and 102. The first asserts that some things occur through chance, even in regard to the first cause (God), while the second asserts that nothing happens through chance. However, a number of the condemnations directly target natural philosophy, as it was understood at the time, by attacking astrology. Rejecting the notion that celestial influences dispose people to differing personalities and gifts, that anyone's health or sickness is dependent upon the locations of heavenly bodies, or even that the stars might indirectly affect an individual's soul, it is clear that Tempier would brook no sympathy toward astrological beliefs. While most modern people don't see astrology as falling within the realm of science, medical doctors and surgeons saw it as an essential field of study and those interested in natural philosophy—such as Roger Bacon and Albert the Great—operated from the assumption that heavenly bodies affected everything from the weather to animalistic impulses—including those of humans—such as lust.

So were the Condemnations of 1277 an attack on science, as the historian of science and physicist Pierre Duhem once thought? As Duhem himself discovered, this incident was nothing of the sort. Although he did note that the condemnations "destroyed certain essential foundations of Peripatetic [meaning, Aristotelian] physics," that turned out to be no bad thing at all. That's because what the condemnations attacked were Aristotelian positions such as the idea that there can be no vacuum. As anyone with even a passing interest in space science now knows, for something Aristotle was convinced could not exist, there certainly is a lot of vacuum in the cosmos. Tempier didn't know that, of course, because no one in the thirteenth century had ever encountered a vacuum. But he and the members of the committee he organized were absolutely convinced that God could create anything he damn well chose to create—including a vacuum. Therefore, they saw Aristotle's statement that vacuum can't exist as a limitation of God's power, which is a definite no-no. Thus, they condemned such statements—which got those with an interest in natural philosophy to thinking about what a vacuum would be like, and how things might move in a vacuum. The result was a great deal of mathematical reasoning that eventually culminated in the French priest and University of Paris professor, Jean Buridan (c. 1295–1363), formulating impetus theory. Buridan saw his theory as a correction to Aristotle rather than something entirely new, and he was building on the work of the Arabic scholar, Avicenna, and the Alexandrian John Philoponus (c. 490–c. 570), but this theory was an important step toward the development of modern physics. In opposition to Aristotle's belief that an object in motion could only stay in motion through the actions of a constant impelling force (such as the air falling in behind an arrow and pushing it forward, in Aristotle's famous "air engine" example), Buridan posited that an object in motion received an impelling force from whatever set it in motion, which would be resisted by the air or other medium it traveled through as well as gravity (though he thought of gravity as the natural tendency of an object containing the element earth to be drawn toward the center of the earth). While not quite Isaac Newton's (1643–1727) notion of inertia, Buridan was getting awfully close.

The point is that there is no evidence that the Condemnations of 1277 did anything to slow down or repress the development of science. Duhem, in fact, came to believe that these condemnations gave birth to modern science. That is almost undoubtedly an exaggeration, in large part because the effects of these condemnations were rather limited, due in no small part to the fact that most people simply ignored them. There's certainly no evidence that anyone ever got into trouble for violating any of the condemnations. However, it does seem that in some limited ways—such as stimulating new avenues of thought in the field of physics—the condemnations had a positive effect on the development of science, rather than retarding its development.

While that positive effect may have been unintended, it's hard to argue that the overall positive effects of the Catholic Church on natural philosophy were anything other than intentional. From Hugh of St Victor (1096–1141), the German monk who described the universe as a machine, suggesting it operated according to natural, mechanical laws that could be teased out by the application of reason, to Robert Grosseteste (1175–1253), the English

Bishop of Lincoln whose work on optics anticipated Newton's in many ways, to Jean Buridan and his development of impetus theory, every major medieval intellectual who had an interest in natural theology was either a priest or a university professor— and all universities were church-run institutions, dedicated to the training of priests. Thus, the church paid, supported, and protected these men, rather than oppressing them and throwing up roadblocks to their progress. Furthermore, these were all men who were driven to study the natural world because of their desire to understand God better. There's no way to put God on a scale or test his (and all mediaeval scholars thought of God as a "he") reactions to stimuli. God cannot be seen or smelled. So throughout the Middle Ages many scholars thought that the best way to understand God was to study the created world.

There's no better example of that way of looking at the world than Albert the Great, who as I mentioned earlier was proclaimed the patron saint of natural scientists in 1941. Albert was a son of the Count of Bollstadt in modern-day Germany who left home to study at Padua and the University of Paris. He eventually earned his doctorate in theology from the University of Paris, where he taught for many years, when he wasn't called away to fulfill other responsibilities, such as acting as the Bishop of Regensberg in Germany. As a doctor of theology, it's important to understand that Albert's overriding goal was to understand God and humankind's relationship to God. And as such he wrote works that would be recognized by a modern person as strictly theological in nature, such as the unfinished—yet still, impressively comprehensive—*Summary of Theology* that he was working on when he died. However, he also wrote many other works that might surprise a modern person, such as a book on zoology and another on geometry. And even when he was writing "pure" theology, Albert would often go off on what a modern person might see as a tangent when discussing a topic such as God's divine power, only to suddenly bring up comets and spend time considering what a comet might be and what its place in the universe is. For this reason, modern scholars sometimes write about Albert's work on zoology or astronomy separate from his theological work, but that's a mistake. For Albert, *all* of his work was theological in nature. God created humans and angels, comets and demons, so any consideration of the natural world is just as much a consideration of God and his divine plan as when writing about salvation or the virgin birth. The distinctions found in modern scholarship are reflective of the pigeon-holed ordering of modern minds, in which questions about God tend to be kept separate from questions about the world. However, Albert wouldn't have understood such distinctions.

So what about the slow pace of scientific developments in the Middle Ages, a slowness that made Carl Sagan feel justified in ignoring this thousand-year long period altogether? Most textbooks on the history of science do little to dispel the notion that this period contributed more or less nothing to the development of science. For example, in my own class I've assigned Frederick Gregory's *Natural Science in Western History*, in part because it does a better job with this period than most—but it assigns only eleven pages to an examination of the medieval period. And it's true that there are few developments one can point to in the Middle Ages as examples of the fruits of scientific thought. There is the occasional technological advance, such as the development of eyeglasses in the thirteenth century, theories such as Grosseteste's ideas about visible light and Buridan's about objects in motion that certainly laid the foundations for or anticipated later scientific thought. And sometimes we find writers like the French doctor of theology and Bishop of Liseux, Nicole Oresme (c. 1320–1382), discussing ideas such as the possible existence of alien worlds that would fit in well with modern cosmological thought. However, overall, the scientific advances are few and far between by modern standards.

That last phrase, though, is key—by modern standards. We should keep in mind that throughout the early centuries of the Middle Ages life was simply too dangerous, uncertain, and impoverished to encourage or support more than a vanishingly small number of scholars. Even as life improved in Europe, with higher crop yields and more safety and security in the kingdoms and city states of the high medieval period, Europe was incredibly poor and life was very uncertain by modern standards. Universities now existed where hundreds of intellectuals and students could be found debating and writing about a host of academic topics, but their numbers were miniscule in comparison to the vast swathes of peasants who lived lives that were all-too-often "nasty, brutish, and short," as the seventeenth century philosopher Thomas Hobbes (1588–1679) put it. Furthermore, just as in Ancient Greece it was prestigious to be a teacher, but none at all to be a developer of new technology much less someone who carried out the experiments, which would have been seen as the sort of nasty manual labor slaves did. In medieval Europe it was prestigious to be a priest, a bishop, or a theologian. It was also prestigious to write books filled with detailed, carefully rendered logical arguments about the way the world works and God's relationship to that world. But there was no prestige associated with carrying out experiments, gathering empirical data about the world, and for that matter the pursuit of natural philosophy as a means in and of itself was far less prestigious than writing about God. The former was the sort of thing a mere master at a university could do, while the latter was the sort of activity reserved to doctors of theology.

Given that context, it's actually rather impressive what the medieval world contributed to the development of science. As Edward Grant argued in his work on *The Foundations of Modern Science in the Middle Ages*, the first and most important medieval contribution was the recovery of Aristotle. His writings on natural philosophy had been lost to Europeans during the early Middle Ages, and if it hadn't been for the dedicated work of translators such as Gerard of Cremona, they would have remained lost, which would have meant Europeans would have had to start from scratch in their efforts to understand the natural world. Furthermore, the development of universities, bringing together the best and brightest minds of Europe while providing a systematic education for many generations of young scholars was a development of incalculable value. Finally, it was very important that scholars such as Jean Buridan began to shake off Aristotle's authority and develop new ideas about the natural world, even if these ideas were incremental in nature and entirely bookish and without the support of empirical evidence. At every step of the way, the church was the primary agent of support for scholarly activity, and the scholars in question were always priests, monks, or friars. Therefore, rather than the Middle Ages being a period of warfare between religion and science, it was most often one of collaboration and support between these two ways of understanding the world.

Further Reading

Bacon, Roger. *Compendium of the Study of Theology.* Thomas Maloney, ed. and trans. Leiden: Brill, 1988.

Colish, Marcia L. *The Medieval Foundations of the Western Intellectual Tradition.* New Haven: Yale, 1999.

Chenu, Marie-Dominique. *Nature, Man, and Society in the Twelfth Century: Essays on New Theological Perspective in the Latin West.* Toronto: The University of Toronto Press, 2013, 2nd printing.

Duhem, Pierre and Roger Ariew. *Medieval Cosmology: Theories of Infinity, Place, Time, Void, and the Plurality of Worlds.* Chicago: The University of Chicago Press, 1987, 2nd printing.

Epstein, Steven A. *The Medieval Discovery of Nature.* Cambridge: Cambridge University Press, 2012.

Grant, Edward. *The Foundations* of Modern Science in the *Middle Ages.* Cambridge: The Cambridge University Press, 1996.

———. *A History of Natural Philosophy: From the Ancient World to the Nineteenth Century.* Cambridge: Cambridge University Press, 2007.

Hendrix, Scott E. *How Albert the Great's Speculum Astronomiae Was Interpreted and Used by Four Centuries of Readers.* Lewiston: Edwin Mellen, 2010.

Lindberg, David. *The Beginnings of Western Science: The European Scientific Tradition in Philosophical, Religious, and Institutional Context, 600 BC–AD 1450.* Chicago: University of Chicago Press, 1992.

———. "Medieval Science and Its Religious Context." In *Constructing Knowledge in the History of Science,* Arnold Thackray, ed. In *Osiris,* 10 (1995): 60–79.

———. "Medieval Science and Christianity." In *Science and Religion: A Historical Introduction,* Gary B. Ferngren ed. Baltimore: Johns Hopkins University Press, 2002, 57–72.

Peters, Edward. *Inquisition.* Berkeley: The University of California Press, 1989.

Wippel, John F. "The Condemnations of 1270 and 1277 at Paris." *The Journal of Medieval and Renaissance Studies.* 7.2 (1977): 169–201.

CHAPTER 3

Medical Knowledge and the Study of the Heavens in the Renaissance

T he Renaissance has certainly had some good publicists. None have been more important than Jakob Burkhardt, the Swiss historian who established the idea of the Renaissance as a distinct historical period in his famed 1860 work on *The Civilization of the Renaissance*. This two-volume study has had an outsized historiographical importance, and it's due in no small part to its influence that history classes typically teach that the Renaissance was a break from the medieval period and a predecessor to the Scientific Revolution, a period which actually overlaps with the Renaissance. Burkhardt saw it as period of explosively new expression, a specialness that extended not just to the art and architecture that so often springs to mind when we think of the Renaissance—he included a section on "Natural Science in Italy" in volume 1 of *The Civilization of the Renaissance*. In this section he praised Renaissance intellectuals for generating the "sort of general interest in the subject [science] which prepares for new pioneers the indispensable groundwork of a favorable predisposition in the public mind," even if they made few scientific advancement. In his view, writers such as Aeneas Sylvius Piccolomini, the man who would become Pope Pius II in

1458 until his death in 1464, inspired an interest in the natural world and the careful observations of that world that would lead directly to the Scientific Revolution, even if they didn't carry out any actual science. And why didn't they? There are many answers to that question, but for Burkhardt, one reason was the inquisition in Italy. He wrote "the Church showed itself . . . actually hostile . . . to genuine science only when a charge of heresy or necromancy was also in question," which has the benefit of being both true and sounding quite promising. However, he goes on to add that such a situation "was often the case." He then notes "it would be interesting to decide . . . whether and in what cases the . . . Inquisitors in Italy were conscious of the falseness of the charges, yet condemned the accused . . . from hatred to natural science, and particularly to experiments." He notes that it is "not easy to prove" that such a thing ever happened, but states, without evidence, that "it doubtless occurred." In other words, according to Burkhardt, the Renaissance was an iconoclastic period, flush with new ideas, that laid the foundations for the development of modern science, but didn't quite make the leap, due in large part to the institutional power of that repressive institution, the Catholic Church. So was he right? Yes and, in a very big way, no. Renaissance ideas and approaches to the world did provide important groundwork for and lead directly to the ideas and practices that would be central to the Scientific Revolution. However, the reason why that revolution didn't occur during the Renaissance (and as we'll see, it actually *began* in the Renaissance, as history is no respecter of neat periodization) had everything to do with social, cultural, and economic forces, and nothing at all to do with the Catholic Church and religious dogma. As with the Middle Ages, the Catholic Church supported the development of scientific approaches to knowledge during the Renaissance.

Let's begin with a few words about the Renaissance in general before we narrow our discussion down to an evaluation of the historical relationship between religion and science in the period. First of all, there's no way to date when the Renaissance began—we have no singular event, like the forced retirement of the last Roman emperor in the west in 476, to stand as a starting point. And since at least 1919, when the Dutch historian Johan Huizinga published a work known most commonly by the English title, *The Waning of the Middle Ages*, historians who specialize in the medieval period tend to think of the Renaissance as simply the late Middle Ages, or at the very least a development of the Middle Ages, rather than an era apart. Needless to say, there are plenty of scholars of the Renaissance who would be quick to disagree, but beyond these dry academic disputes it's important to point out that no term like "Middle Ages" or "medieval" was used until the very late Renaissance, when the Latin term *media tempestas* (middle time) first made its appearance in 1469. The term gave a name to the idea held universally by Renaissance intellectuals that the centuries between the collapse of Roman power in the West and the Renaissance were in the "middle" of two glorious periods, and was itself devoid of anything of merit. As for the term "Renaissance," it would not be used until 1858, when the French historian Jules Michelet coined it to refer to a "rebirth"—which is what the word *renaissance* means in French—of classical creativity and knowledge, only two years before Jakob Burkhardt took the idea and ran with it. Nevertheless, intellectuals living in the Renaissance felt themselves to be alive in a time that differed from, and was superior to, the era that came before it.

In many ways the Florentine poet, Francesco Petrarca (1304–1374), commonly known simply as Petrarch, lived a life that marked the beginning of the Renaissance and the attitude of Renaissance scholars toward the Middle Ages. Born the son of a successful merchant, notary public, and minor politician in the Florentine Republic—Italy wouldn't coalesce into a single nation until 1861, and even then the nation wouldn't include Rome until 1870—, Petrarch had from his youth the means to pursue an education and the influence to drive him to do so. Unlike modern notaries, in fourteenth century Italy they were minor lawyers who handled many of the day-to-day activities that require such a professional, such as drawing up wills and contracts. Such men had to have a moderate level of education, as did merchants of the day, and there was plenty of work for notaries drawing up contracts and the like. Italy had long been the center for trade between West and East. Merchants from city-states such as Florence and Venice traded with Muslim-controlled lands such as Egypt, what was left of the Byzantine Empire in Asia Minor, and even traveled as far as China in search of trade. Therefore, Italy was wealthier and more urbanized than any other part of Europe, and had avenues of education, such as the so-called Abaca schools, which provided a grounding in literacy and numeracy for a relatively large number of people with a professional need of such skills. Furthermore, even after the ravages of centuries, Italy possessed a large number of monuments both large and small built by the Roman Empire that had been born there. Thus, Petrarch grew up in a home that had more books than most, surrounded by men who valued learning, and confronted on a daily basis by the massive works that the Romans had left behind.

Like fathers throughout time, Petrarch's father wanted his son to attain more and do better in his life than he had. Therefore, he pushed his son to attend university and pursue a doctorate in law, which would allow him to hold a far more prestigious position than that of a notary. Petrarch dutifully attended first the University of Montpellier and then the University of Bologna. However, in his seven years at these two institutions, he could never "face making a merchandise of [his] mind," as he put it, for he saw the practice of law as one of selling justice. Therefore, he focused on his true love, literature. As anyone interested in making a writing a career should do, he read voraciously while honing his own skills as a writer by producing poems or narratives in the form of letters to friends and family across Italy, and eventually across Europe. It was during this period that Petrarch fell fully in love with the writers of the Roman Republic (a period from roughly 509 to 27 BCE). Christian writers had long held that the period of Classical Antiquity was an age of darkness because it lacked Christianity, but Petrarch would come to reverse that and write that, during the Roman period, "Amidst the errors there shone forth men of genius; no less keen were their eyes, although they were surrounded by darkness and dense gloom." In other words, in spite of the fact that they lacked the advantage of Christianity, the Romans were men of genius. It was the men of the Middle Ages who lived in true darkness, because they were mired in barbarisms and lacked this genius. Thus, it is to Petrarch that we owe the idea that the Middle Ages were "Dark Ages," lacking in intellectual and artistic development. Similarly, the term "Gothic," first applied to the soaring architectural style developed by an unknown medieval architect and visible in buildings such as the Cathedral of Notre Dame, developed during the Renaissance as a term of abuse. The Goths were barbarians who had sacked and eventually overrun Rome, thus by referring to medieval architectural and artistic styles as "Gothic," intellectuals of the Renaissance meant to imply that it had destroyed the beautiful contributions of the Romans.

Those contributions included cities and infrastructure in Italy that never quite collapsed along with Roman power in the West. Therefore, the city-states of Italy needed less recovery time during the Middle Ages, and by the fourteenth century they were growing explosively in wealth through trade with the East. As they did so, they also grew increasingly competitive with one another. This competition sometimes broke out into outright war, but all in all war is bad for business. Therefore, the governments and leading families of Italian city-states such as Florence, Venice, and even the papal states—which from 1307 to 1377 were actually governed by popes who lived in the French town of Avignon, and from 1379 to 1415 there would be both a pope at Avignon and one at Rome—all sought other, less destructive, forms of competition. This included hiring architects to design buildings more elaborate and beautiful than those next door and artists to decorate those buildings and construct statues and other ornaments to beautify the cities, while often sending a message to opponents. For example, the Opera del Duomo of the Cathedral of Florence originally hired three different artists beginning in 1410 to create the first in a series of Old Testament figures in marble. For various reasons, none of these men finished the job, so in 1501 the youthful but already famous Michelangelo (1475–1564) received the commission to complete the work. When he had finished the fourteen-foot tall masterpiece known as *The David*, the city council decided to place it in the Piazza della Signoria, the political heart of the city. Carrying a slingshot that is almost invisible, the expression on the statue is of supreme cleverness and self-confidence—the sort of abilities that would allow the much smaller David to defeat his larger opponent, Goliath. That's why the Florentine city council placed the statue in the heart of the city and in sight of the citizenry, facing in the direction of Florence's great rival and sometimes outright enemy, the larger and more powerful city-state of Venice.

It should be noted that the practice of supporting artists and scholars as a form of competition wasn't restricted to the level of city governments and guilds. Italy had many families who had earned wealth through trade or the banking that trade necessitates, but who lacked the social position to allow them a level of prestige equal to their wealth. One way these families—families like the Medici, a banking dynasty who would come to more or less rule Florence from behind the scenes by the end of the fourteenth century, and who would produce four popes—would increase their prestige was by using their wealth to secure marriages to members of older, landed, but not always wealthy, nobility. Another way was by acting as patrons to artists, poets, scholars, and other intellectuals who would produce works dedicated to their patrons, who could then bask in the reflected glow of their genius—while also showing off their wealth by demonstrating that they had money to burn. Therefore, talented individuals such as Petrarch who managed to attract the attention of the wealthy and the culturally influential could work at their craft full time through the financial support of their patrons. A certain level of wealth is necessary for a culture to produce scholars and artists. One doesn't necessarily follow the other—sometimes wealthy societies produce reality television, for example, instead of works with cultural merit—but in spite of the trope of the starving artist, those who truly starve are hardly capable of work with cultural or intellectual merit.

We should take note of one more factor before we move on. This vigorous competition to attain status while one-upping one's neighbors led to a fair amount of wealth flowing into the hands of artists such as sculptors and painters, many of whom became famous in their own right. Poets such as Petrarch became famous—in 1341 he became the second person to be honored with the revived classical tradition of being crowned with laurel leaves as a poet laureate—a tradition that went back to the ancient world, but now so did sculptors such as Donatello (1386 and 1466), Michelangelo, and many others. Throughout the Middle Ages it's rare indeed to find a sculpture or painter who can be identified by name. Throughout the same period intellectuals publicized the names of writers such as the Greek poet Homer (c. 750 BCE, assuming that "his" work wasn't actually a conglomeration of poetry from numerous authors) and the Roman statesman and philosopher, Cicero (106–43 BCE). However, the identities of Greek artists such as the fifth century BCE Athenian sculptor, Kritios or the second century BCE Roman painter, Publius Aelius Fortunatus were lost. The reason why is simple—during the Middle Ages, as we saw in chapter 2, there was no prestige associated with sculpture, architecture, painting, or other forms of art associate with manual labor. But in the Renaissance that changed, as artists played a significant social role and received benefits (or at least the best and the luckiest artists did) in the form of the wealth and prestige that went along with that role. During the Renaissance, it once again became cool to work with your hands, so long as the work you were doing was seen to have cultural and intellectual merit.

Thanks in part to the habit the Renaissance wealthy had of supporting artists, Petrarch would go on to have a long and productive career in the public eye. Not only did he write large numbers of poems, often in the Florentine dialect of Italian, but he also sought out and uncovered rare manuscripts, such as the letters of the Roman statesman, Cicero, showing that Cicero had been far more than the philosopher the Middle Ages thought him to be—he had in fact been the most important member of the Senate during a crucial period when the Roman Republic was coming to an end. Petrarch sought out such manuscripts, because he insisted that if anyone is to truly know a writer like Cicero, one must take an *ad fontes*, "back to the source," approach and bypass the works *about* that author, or containing snippets of the author's writings, and instead read the author directly in his or her original language. This approach became a key component of Renaissance humanism, which in its essence was built around an overwhelming confidence in human capability and a belief that direct study of the Ancient Greeks and Romans will allow scholars to create something newer and better, built on the foundation the ancients laid. Thus, although Petrarch never actually learned to read Greek himself, he encouraged others to do so, and within a century scholars of Greek, Hebrew, and Arabic would begin popping up first in Italy, then France, Germany, and every other part of Europe as the Renaissance spread northward. It's not clear how much Petrarch understood his role in the development of the Renaissance—though he wasn't given to undue humility—but by his death in 1374 the humanist movement had already taken root.

However, we should not read that as a turning away from religion. In 1888, the English poet and cultural historian, John Addington Symonds, wrote in volume 1 of his study of the *Renaissance in Italy* that "what the word Renaissance really means is new birth to liberty—the spirit of mankind recovering consciousness and the power of self-determination . . . liberating the reason in science and the conscience in religion" (a quote that's often misattributed to Burkhardt, incidentally). That seems to suggest a turning away from religion, or at least a new direction for religion. Students often hear the word so central to Renaissance intellectual life, "humanism," as some sort of antonym to "religious," due to the modern association of the term with "secular" in some American political circles. It's true that Petrarch, in line with many other humanists, wanted religion to be more personal and less intellectualized than scholastic philosophy promoted. Like any good humanist he argued for an *ad fontes* approach to religion, as with everything else—he advocated reading and contemplating the bible for oneself, and suggested that reading it in the original languages would be best, though Petrarch never attained the knowledge of Greek and Hebrew this would have required. As with all humanists, he believed that human beings have enormous potential—a potential given by God, that He wants all humans to develop. However, none of this is really at variance with medieval religious thought (other than the rejection of scholasticism, that is); Albert the Great and Thomas Aquinas both wrote frequently of humankind's free will, it's importance, and the capabilities it provided for humans to strive for greatness. Furthermore, as Petrarch stated in numerous letters, but nowhere more forcefully than in his meditation upon the monastic life, *On Religious Leisure,* he firmly believed the religious vision of life to be superior to anything secular culture had to offer. While he was critical of some of the abuses that befell monasteries, he viewed the quiet solitude of a monastic setting as the perfect atmosphere for the contemplative, Christian life. We should remember that Petrarch spent much of his life as a church canon, in religious orders.

In part, Petrarch's decision to become a canon could be described as a means of getting a guaranteed—albeit limited—income. However, given his expressed personal piety, we shouldn't write it off as simply that. And he had plenty of other sources of income, thanks to the wealthy Italians who wanted enhance their own prestige by supporting him. As mentioned, such support went to artists and scholars of all types, and while the work these men did enhanced the prestige of their patrons, the incomes they received and the fame of the work they completed for wealthy and famous patrons provided a personal boost to their own prestige. This would be important for the development of science, in part because of the enhanced prestige of arts such as sculpture and painting mentioned above. Artists, infused with humanist impulse of the Renaissance, glorified the human body in ways that most of their medieval predecessors wouldn't have understood—not because of any religious prohibitions, but simply because of a difference in outlook. Already by the mid-thirteenth century in Italy we find nude representations of the human body based on classical models, but by the mid-fifteenth century we find nudes everywhere in art—and not only nudes, but artistic representations that are obviously drawn from close examination of the human body.

The Catholic Church as an institution not only didn't oppose nude presentation of the human body, it promoted it. During the Middle Ages it's true that only Adam and Eve were routinely portrayed nude, in line with Genesis' description of them before the fall, but during that period art lacked the prestige or the funding necessary for the existence of more than a handful of professional artists. Furthermore, the medieval attitude toward manual labor—that working with one's hands was a thing reserved for the lower classes—precluded the sort of people from engaging in art who had the leisure to sit around looking closely at naked people for the purpose of creating art about them. But the fifteenth-century representations of the infant Jesus were frequently not only nude, but done in a way that put his genitals on clear display, as a means of emphasizing that he was God made man. By the sixteenth century, artistic representations of the Virgin Mary sometimes depicted her with her breasts exposed to emphasize her role as a mother, and of course between 1508 and 1512 Michelangelo covered the ceiling of the Sistine Chapel with numerous nudes, with the approval and in the pay of Pope Julius II (r. 1503–1513).

The artistic examination of the human body went beyond simply having a model sit for the artist. As Giorgio Vasari wrote in his *Lives of the Artists*, the Florentine sculptor and painter, Antonio Pollaiuolo (1431/32–1498) became the "first master to skin many human bodies in order to investigate the muscles and understand the nude in a more modern way." By the fifteenth century, patrons often wanted an artist not only to be knowledgeable about dissection practices, but to have also carried out dissections to learn how muscles are put together beneath the skin, how the bones act as a superstructure for the body, and how the joints of the limbs move. For example, the Florentine sculptor Baccio Bandinelli (1493–1560) ran an academy for young artists while seeking out patronage as vigorously as any of his contemporaries, and it was in that pursuit that he told a duke who was considering commissioning a work from him: "I will show you that I know how to dissect the brain, and also living men, as I have dissected dead ones to learn my art."

In spite of statements made by those such as Sarah Jane Boss, the director of the Center for Marian Studies at Roehampton University, in her study of *Mary* that "for hundreds of years the Catholic Church forbade the anatomical dissection of human bodies, or permitted it only under the strictest ecclesiastical supervision," the Catholic Church didn't forbid dissection. And the "strict" ecclesiastical control that Boss refers to was actually a prohibition against performing *public* dissections without ecclesiastical permission. Confusion occurs because in 1163 the Council of Tours issued a prohibition against clergy or monks shedding blood, which some took to mean that such people could not practice medicine. However, this statement was aimed at stopping unqualified people from practicing surgery, and it had nothing to do with dissection. Another point of confusion occurs because in 1299 Pope Boniface VIII (r. 1294–1303) made a pronouncement against the practice of dismembering the bodies of dead crusaders and boiling the parts in order to remove the flesh (a process used to make it easier to return their bones to their homelands). However, this bull had nothing to do with dissections or autopsies, which was how it was understood in Italy. However, as Katherine Park writes in her consideration of the myth of "The Church's Prohibition of Human Dissection," a small number of northern Europeans expressed concern that this bull might prohibit dissection. Nevertheless, as she explains, there is no evidence of any anatomist or surgeon ever being prosecuted for dissecting human bodies, nor is there evidence of any church official ever refusing to give permission for a public dissection. The real bar to dissections at medieval universities was twofold. First and foremost, from the time of the Romans on there were cultural taboos about cutting up bodies. Among the Romans, touching dead bodies was considered unclean, and among medieval people it was considered disrespectful to the dead—not to mention that it put the body into a condition where public observance of the corpse during a period of pre-burial grieving (which was a common practice in regions such as Italy) wasn't possible. These cultural taboos about cutting up dead bodies were only magnified by the fact that in the Middle Ages those who were of the intellectual class simply found any sort of manual labor distasteful and beneath them.

That aversion to manual labor—and labor was defined rather broadly, to involve any sort of manual activity beyond hunting or war—could manifest itself in ways that seem very odd to a modern person. Today surgeons see themselves—and are normally seen by others—as an elite class among physicians, but nothing could have been further from the truth in the Middle Ages and Renaissance. It was not until the fourteenth century that the works of learned men such as Lanfranco of Milan (c. 1250–1306) and Henry de Mondeville (c. 1260–1316) began to have an impact in Italy and France, respectively. These men were surgeons who had received a university education, which was very unusual for the day. They applied this education to writing scholastic works that applied the medical theories of the second century Roman author, Galen, and Aristotle's logic and natural philosophy to the practice of surgery, thereby making a case that this was a discipline every bit as learned as that of physician. Up until this point, and for that matter well into the fourteenth century, physicians would never actually touch a patient, instead diagnosing his or her ailments by an examination of urine and the casting of a horoscope in conjunction with listening to a description of symptoms. If any cutting or bloodletting—a very common treatment—was to be done, the physician would refer the patient to a much-lower-status surgeons. Slowly, during the fourteenth century, surgeons began to professionalize on the continent (though it would be the fifteenth century before professionalization occurred in England), as surgeons' guilds set standards for entry into the profession that included a university education and a mastery of Aristotle and Galen. Nevertheless, it would be quite some time before the status of surgeons would equal that of physicians, due in no small part to the low opinion the upper classes had of manual labor. Thus, it was culture, not religious prohibition, which kept dissections from being a regular part of the curriculum at medieval universities.

However, during the Renaissance cultural changes occurred, and it was largely because of the important social functions art played and the subsequent rise in influence of artists. By 1400, anatomical dissections were a regular part of medical instruction for physicians who studied at many universities. Over the course of the century, it became common for medical professors to read from Mondino dei Luzzi's (c. 1270–1326) *Anatomy* while directing an assistant who would perform a dissection in front of medical students (physicians still didn't cut). Mondino's *Anatomy* included an outline for dissectors to follow, based on his own experiences using public dissections to teach anatomy at the University of Bologna, the first of which he carried out in 1313, with ecclesiastical approval. And while it might sound odd to a modern person that approval from the church was necessary for public dissections, we should keep in mind that all universities were run by the church and, at least in theory, university students were priest-in-training, even if many would never actually perform such a role. The only requirement in order to be able to carry out public dissections was that one had to be a qualified teacher—or an assistant to a teacher—and performing the dissection for educational purposes. So, rather than the requirement that any public dissection receive ecclesiastical approval representing the act of a repressive church, it was instead a regulatory act meant to restrict the teaching of medicine to qualified medical schools. There are similar regulations today about the use of cadavers, though now it is the state rather than the church that restricts and regulates how and under what conditions dissections can be performed.

Among those who think the Catholic Church repressed the development of science, the example of Andreas Vesalius (1514–1564) is sometimes held up. Vesalius was born in Brussels to an apothecary who was the son and grandson of physicians, so it made sense for the younger Vesalius to be drawn to medicine. To that end, he attended university at Louvain, Paris, and Padua at an exciting time. The humanist search for classical works had uncovered early examples of Galen's anatomical work, *On the Use of the Parts,* and by 1500 a fine Latin translation drawn from the earliest textual examples was available. By 1523 a Latin translation of his physiological work, *On the Natural Faculties* became available. Finally, in 1531 a medical humanist at the University of Paris named Johannes Guinther (1487–1574) published an edition of a recently discovered work by Galen, *On Anatomical Procedures,* along with his commentary on the text. While at Padua, Vesalius had the good fortune to study under Guinther, so he gained the most cutting-edge education in anatomy available.

After obtaining his doctoral degree in 1537, Vesalius personally performed public dissections for medical students. His method was different than that of his teacher, Guinther, who had directed dissections performed by an assistant—which meant Vesalius saw more, and understood the smallest, most intricate details, of human anatomy with greater precision than his master had. He even provided private dissections for advanced students. In order to help these students with their studies, he made a series of anatomical charts and published six of them in 1538, when he learned that someone else had published a couple without permission. As an experienced dissector and close observer, Vesalius realized something: Galen was often wrong, and wrong in ways that could only indicate that he had performed his own dissections on animals instead of humans, which is exactly what had happened. As discussed in chapter 1, Galen had worked in Rome where dissection was forbidden, not because of any religious restriction—Galen lived in second century Rome where Christianity was outlawed—but because it was considered incredibly disrespectful to cut up a human body. Vesalius was the first anatomist to recognize the mistakes Galen had made; humans have a strong tendency to see what they expect to see, and Galen was considered *the* authority in the field of anatomy. This was especially true in sixteenth-century Europe, where natural philosophers were having something of a love affair with the Roman anatomist given the recent spate of translations and discoveries of ancient Galenic manuscripts.

Vesalius was ambitious, and he yearned for recognition not only as an anatomist, but also as a philosopher, which was far more prestigious in the sixteenth century. Therefore, he used Galen as his philosophical model, but expanded upon him, correcting the details of his anatomy where necessary but always adhering to Galen's philosophical principles when he wrote his own work on anatomy, *On the Fabric of the Human Body*, which he published in 1543. Most importantly, this work contained 273 beautifully detailed illustrations of human anatomy. The invention of printing in the mid-fifteenth century made accurate reproduction of these drawings possible, and soon physicians across Europe were seeking out copies of *On the Fabric*. That was Vesalius' intended audience, as students were unlikely to be able to afford the beautifully illustrated work. In hopes of gaining patronage, he dedicated it to the Holy Roman Emperor Charles V (r. 1519–1556). The dedication had the desired effect, for it drew Charles' attention, and he soon offered Vesalius a position as his court physician. Saying goodbye to academia, Vesalius became the personal physician to first Charles V, then his son Philip, who ruled Spain as Philip II from 1556–1598.

According to Andrew Dickson White, "[i]n the search for real knowledge [Vesalius] risked the most terrible dangers, and especially the charge of sacrilege, founded upon the teachings of the Church for ages." In passages that read more like a historical novel than sober history, White paints Vesalius as a hero willing to brave "ecclesiastical censure" and the forces of the inquisition in order to tell the truth, bravely performing dissections in spite of papal prohibitions on such activities. Now, it is true that Vesalius found himself intermittently embroiled in controversy, and there is a contemporary report of the papal inquisition prosecuting Vesalius. However, the French diplomat Hubert Languet (1518–1581) is the source of the report of Vesalius' troubles with the inquisition, and as a committed Protestant, (France endured decades of civil war between Calvinist Protestants and Catholics from 1562–1598) he spread this story not as an unbiased reporter of facts, but as a means of making one of the great defenders of Catholicism, Phillip II, look bad. Given the sensationalist nature of Languet's account—that Vesalius performed an autopsy upon a Spanish nobleman whose heart was still beating—it's likely that the only reason anyone accepted him at his word was because it fit with the preconceptions that Protestants had of Catholics during this period. The controversy, however, had nothing to do with Vesalius' dissections. He had been openly doing such work for decades, and his professors in medical school had done so as well—dissections and autopsies were common in sixteenth century Europe, and a standard part of every medical student's education.

The controversies Vesalius found himself embroiled in stemmed from a very different cause; he was upsetting the academic apple cart. European physicians, surgeons, and anatomists were widely committed to Galenic theory and Galen's anatomical notions. Everyone in Europe who studied anatomy did so by reading Galen, and academics across Europe had invested a great deal of energy teaching his ideas, writing about them, publishing commentaries on Galen, and otherwise building their reputations on his works. To suddenly be confronted by some upstart from the fringes of Europe who said Galen was wrong . . . frankly, it's surprising the furor of opposition wasn't stronger. Philosophers of science such as Pierre Bourdieu and Thomas S. Kuhn have stressed how resistant scientific professionals are to new ideas that undercut the work they've built their own reputations on, a fact that is as true today as it was in the sixteenth century.

Thus, when Charles V convened a commission to investigate the religious implications of Vesalius' work in 1551, this was a means of countering his detractors who used unfounded and unsubstantiated attacks on Vesalius' religious beliefs as a form of mudslinging to attack someone whose work undermined their academic publications, not as an act of persecution. The commission cleared Vesalius, and this event had nothing to do with Vesalius' later decision to go on pilgrimage to Jerusalem in 1564, in spite of White's assertion that the religious authorities drove Vesalius out of Spain and forced him to become "a wanderer" "on a pilgrimage to the Holy Land, apparently undertaken to atone for his sin." Devout Catholics sometimes went on pilgrimage to Jerusalem in the sixteenth century—as has been true of centuries of Christians before and since. Unfortunately, sea travel was inherently unsafe in the sixteenth century, and a storm shipwrecked Vesalius on the coast of Corfu, where he died at the age of forty-two.

As a closing note in our consideration of Vesalius, White includes a very interesting paragraph in his discussion of Vesalius' problems, which is instructive enough to be worthy of quoting at length. White wrote:

> Throughout the Middle Ages it was believed that there exists in man a bone imponderable, incorruptible, incombustible, the necessary nucleus of the resurrection body. Belief in a resurrection of the physical body, despite St. Paul's Epistle to the Corinthians, had been incorporated into the formula evolved during the early Christian centuries and known as the Apostles' Creed, and was held throughout Christendom, "always, everywhere, and by all." This hypothetical bone was therefore held in great veneration, and many anatomists sought to discover it; but Vesalius, revealing so much else, did not find it. He contented himself with saying that he left the question regarding the existence of such a bone to the theologians. He could not lie; he did not wish to fight the Inquisition; and thus he fell under suspicion.

What is most interesting about this passage is that there is no evidence that medieval people held any such belief. I have no idea where White got his information, Vesalius certainly didn't write that he would leave "the question regarding the existence of such a bone to the theologians," and no inquisitor ever put any suspicion he might have had about Vesalius regarding such a bone into any writing that has yet been recovered. Yet White isn't the only one to refer to such a bone. To provide only one example, Howard W. Haggard (1891–1959), the first director of the Yale Laboratory of Applied Physiology and author of numerous works on the history of medicine, mentions it as one of the many "Galenic myths" that Vesalius overthrew in his book *From Medicine Man to Doctor: The Story of the Science of Healing.* He was likely relying on White as a source, or perhaps using a source influenced by White.

What seems to have happened here is a bit of late Victorian medievalism gone awry. During the nineteenth and early twentieth century, there was an intense fascination with the Middle Ages, as evidenced by writers such as Sir Walter Scott (1771–1832) and the painter John William Waterhouse (1849–1917). The Middle Ages these artists depicted was highly romanticized and chock full of elements either made up out of whole cloth in order to make for a better story (or painting), or based on myths people held about the quaint beliefs of the Middle Ages. The myth of the resurrection bone seems to be such a case and may have been developed by reference to Jewish beliefs. As Geoffrey W. Dennis discusses in *The Encyclopedia of Jewish Myth, Magic and Mysticism* (s.v. "Bone"), "the Midrash speaks of a bone called the luz that, like the soul, is indestructible. It is from this bone that God will resurrect the body." Rabbinical writings discuss this belief, but it is not one found among Christians of the medieval or early modern periods. However, once White wrote so confidently about the resurrection bone, the idea that medieval people held this belief, which Vesalius overthrew, pops up from time to time in popular histories and even scientific writings, usually as a means of reinforcing how deluded medieval intellectuals were. It's unknown where White got the idea that Christians held this belief, as earlier nineteenth century writings about the resurrection bone always accurately referred to it as a Jewish belief. One thing is certain though: no religious authority opposed Vesalius because he didn't find the resurrection bone, since this was not a belief Christians held, and the number of sixteenth-century Christians who had even heard of the concept must have been miniscule.

However, it certainly wasn't only in the area of the arts that the Renaissance had an impact in ways that would affect the history of science. As mentioned above, humanist scholars spread out across Europe, searching monastic libraries and other repositories of ancient texts in hopes of finding unknown Greek or Roman works, or earlier examples of such works than those that previously used by scholars in order to check the accuracy of these works. In this way a number of newer and better versions of Aristotle's works were produced, but more importantly for our topic, there was an avid recovery of Plato and the third-century Platonic writer, Plotinus. This recovery was facilitated by a trickle of Greek-speaking scholars who came to Italy in the late fourteenth and early fifteenth century as the Catholic Church negotiated with the Greek Orthodox Church, which had split away in 1054, for a return to the Catholic fold. Few in the Orthodox tradition found this prospect appealing, but as the Ottoman Empire increasingly encroached upon Byzantine territory, many among the Greeks came to see reconciliation with the Catholics as a necessity in order to gain the military assistance that could allow for Byzantine survival. Alas, such a reconciliation proved impossible, and in 1453 the capital of the Byzantine Empire, Constantinople, fell to the Ottomans. In spite of Ottoman tolerance of Christians, many high-placed officials within the Orthodox Church were frightened at the prospect of Muslim rule and fled the city, bringing Greek manuscripts with them. These highly educated men needed jobs in Italy, and what better job for a well-educated man in a land eager to learn Greek than that of teacher? As a result, though the fall of Constantinople didn't start the Renaissance, which was already well under way at the time of the fall, it certainly gave it an infusion of energy.

As the Greek language and manuscripts began to become more accessible, this (along with Arabic) became the language of prestige among humanists. Thus, merchants and bankers eager to enhance their prestige began to support scholars who knew Greek and the Platonic and Neoplatonic philosophy that so fired the imagination of Renaissance humanists. None of these patrons were more important than the Florentine banker Cosimo d'Medici (1389–1464). As the head of the wealthiest banking house in Europe, with branches as far afield as London, which was deeply involved in both papal and imperial politics, Cosimo had immense wealth, but lacked the prestige and respect that would have come from noble birth or the education Cosimo had never had time to attain. This prestige and respect was particularly important, because Cosimo was maneuvering behind the scenes in Florence to gain greater political power. These maneuvers would establish the Medici as not only the de facto rulers of the Florentine Republic for centuries to come, but also gain the family entry to papal influence—there would be three Medici popes in the sixteenth and early seventeenth century—the hereditary duke-ship of Milan, and through marriage the family would establish close ties with the king of France. One part of this maneuvering was the patronage of artists and scholars that would allow the Medici to be seen as cultured and worthy of respect, instead of as grubby moneylenders.

This patronage took many forms, from supporting archi-tects such as Michelozzo di Bartolomeo Michelozzi (1396–1472) to sculptors such as Donatello (1386–1466). Perhaps most importantly, Cosimo financed the construction of the first pub-lic library in Florence in 1444—a beautiful structure, designed by Michelozzi, which is still a pleasurable place to conduct research—and provided financial support for scholars to work there, while letting it be known far and wide that he would pay well for Greek manuscripts. This led scholars such as Marsilio Ficino (1433–1499) to take up residence in Florence and apply themselves to the translation and analysis of Plato and Platonic writers. Although not ordained as a priest until he was forty, Ficino was a deeply religious man throughout his life. He was also keen to demonstrate the ways in which Platonic thought was consistent with Christian thinking, as best evidenced by his mas-terpiece, *Platonic Theology: On the Immortality of the Soul*, which he worked on throughout the 1460s and completed in 1474. With Cosimo's financial support and encouragement, Ficino translated the extant works of Plato into Latin for the first time in history, a project of incalculable importance as it made Plato accessible to the European intellectual elite, many of whom eagerly embraced Platonism.

Just as important for our current study, though, is the proj-ect Cosimo interrupted Ficino's work on Plato with in 1462. In 1460, a monk named Leonardo da Pistoia (born around 1428, and not to be confused with the painter of the same name) brought a Greek manuscript to Cosimo, which he had acquired in Macedonia. Cosimo placed the manuscript in his library, and two years later asked Ficino to translate what came to be known as the *Corpus Hermeticum* when he completed the project in 1463. Ficino thought the fourteen-book work was the missing link of

Platonic scholarship, because he was convinced that Plato's works represented a truly ancient form of knowledge with direct ties to Christianity. In the works of the philosopher he thought of as Hermes Trismegistus (Hermes the Thrice Great), Ficino believed he had found proof of a direct relationship between Plato's philosophy and Christianity. Ficino believed that Hermes was an ancient Egyptian philosopher, whose works had been passed down to Moses, and through him had been secretly transmitted to both Christianity and Platonic philosophy. Thus, the two belief systems had a common source. This belief made sense in light of Platonic elements found in the bible, such as John 1:1, "In the beginning there was the Word, and the Word was with God." In the original Greek, word is *logos*, and this statement is almost a direct quote from Plotinus; the book of John shows Neoplatonic elements throughout, and Paul's writings also show signs of Platonic influence. However, instead of this influence flowing from Hermes through Moses and then to both Plato and Christianity, it's actually the result of Hellenistic influences on the Jewish community in Palestine. But as far as Ficino was concerned, the *Corpus Hermeticum* provided evidence of the hidden transmission of Hermetic philosophy through the ages. It would be much later that scholars would realize that there was no "Hermes," and these works aren't nearly as ancient as Ficino thought. Instead, they are compilations of works produced in Alexandria in the second and third century.

However, the historical details of Hermetic philosophy—and the fact that there had never been a scholar known as Hermes Trismegistus—were unknown in the fifteenth century, and when the *Corpus Hermeticum* rolled off the printing presses, scholars and natural philosophers across Europe eagerly seized on it. It was hot, trendy, and intellectually exciting to read Plato and Hermes, and both had tremendous influences on scholars across Europe. One such scholar was the son of a Polish merchant from the town of Torun in the Crown of the Kingdom of Poland, Mikołaj Kopernik, most commonly known by the Latinized version of his surname, Copernicus (1473–1543). Even by the standards of the fifteenth century his family was notably religious; not only would

Copernicus enter a religious vocation (becoming a church canon), but his brother would become an Augustinian canon and his sister a Benedictine nun. When Copernicus was only ten years old, his father died, and his uncle, Lucas Watzenrode the Younger (1447–1512), took Copernicus under his wing. Watzenrode was a former schoolmaster at St. John's school in Torun and a humanist with extensive contacts among scholars of eastern Europe, and in 1489 he would become the Prince-Bishop of the province of Warmia, It's likely that Copernicus attended St. John's school and had his first introduction to humanist thought—which included an emphasis on Plato—from his uncle and his uncle's friends, before going on to study at the University of Krakow in 1491. He showed a particular facility with math and astronomy, interests he would take with him when he transferred to the University of Bologna in 1496 in order to study church law.

Humanism was deeply embedded in the curriculum of the University of Bologna, in all areas of instruction, including astronomy. Either because of his natural abilities or his family connections, Copernicus soon gained the attention of a member of the arts faculty who specialized in astronomy, Domenico Maria da Novara (1454–1504). Novara had himself studied under the most famous astronomer and astrologer (the two fields were completely one and the same in medieval and Renaissance Europe) of the day, Johann Muller, a Bavarian scholar best known as Regiomontanus (1436–1476). Regiomontanus was a humanist who was fluent in Greek, most famous for making the second-century Alexandrian scholar Ptolemy's work on astronomy known as the *Almagest* more accessible to European scholars. He completed his *Epitome of the Almagest* in the early 1460s, though it was not printed until 1496. The *Epitome* updated Ptolemy's geocentric (Earth-centered) astronomical system by including astronomical observations and calculations made since Ptolemy's time, while also clarifying it for ease of use by working astronomers. Studying under Novara, Copernicus was more interested in humanistic studies and astronomy than church law (though he would eventually earn a doctorate in law in 1503); he learned Greek, humanist techniques of scholarship, and a great deal of astronomy (as well as a fair bit of medicine, which was closely connected to astronomy during this period)—and he developed an intense interest in Hermetic philosophy. Thanks to Ficino's translation of the *Corpus Hermeticum*, this Neoplatonic philosophy was all the rage at universities such as Bologna where humanism was dominant, and students read it, talked it over, and argued about it incessantly. Thus, it was at Bologna that Copernicus became fully acquainted with both the Ptolemaic system and Hermetic philosophy, and that combination would prove to be very important.

Copernicus' understanding of the Ptolemaic system was important because by the fifteenth century, it was a system in crisis. There were two major problems with it, one theoretical and one practical, and frankly the theoretical one was the most troubling from Copernicus' point of view. As a deeply committed Christian who was an adherent of Hermetic Neoplatonic philosophy, Copernicus believed that God had imbued the universe with rational order. God had created the natural laws that kept the universe functioning, and since God was both wholly perfect and wholly good, those laws should be as perfectly rational as He is. However, in order to make his system work, Ptolemy had introduced a mathematical concept known as the equant point. This is an imaginary point placed opposite the Earth, and both the point and Earth were equidistant from a point between them known as the deferent. It was obvious to the ancients through naked-eye observation of the moon, sun, and other planets (the word "planet" comes from a Greek term meaning "wandering star," so anything that appears to be star like in the heavens, but moves, was a planet so far as the Greeks were concerned) orbit the Earth. Think about it—the moon and sun certainly look as if they are circling the planet, and so do the visible planets if you watch them move across the sky. All observable evidence supports a geocentric rather than a heliocentric model of the solar system, which is why no more than a handful of pre-modern astronomers ever gave any thought to a heliocentric system. However, the Greeks also believed the heavens to be the realm of perfection and assumed that everything in the heavens moved in perfect circles, as they thought of the circle as the perfect geometric shape. But observations of planetary motion didn't exactly match a geocentric model in which planets move in perfect circles, especially since planets such as Mars and Mercury sometimes appear to speed up and slow down, and even go backwards, when observed from the Earth. Therefore, Ptolemy introduced the equant point and epicycles (smaller circles made around an imaginary point on the larger circle that is the orbit of the planet) in order to make his mathematical models match his observations. And while these mathematical concepts made for some complicated math, for the most part they allowed astronomers to predict planetary motion with a high degree of accuracy.

However, in spite of the accuracy of the system, which seemed to suggest it was fundamentally correct, the introduction of the equant point meant that the perceived acceleration and deceleration of certain planets wasn't simply an artifact of observation; they *actually* sped up and slowed down, depending upon where they were in their orbit. As Copernicus would later state in the dedication to his *On the Revolutions of the Heavenly Orbs,*

> when I had long considered this lack of certitude in the
> mathematical tradition concerning the composition of
> motion of the spheres of the world [by which he means
> the entire universe, not just the Earth], I began to be
> annoyed that no more certain theory of the motions of the
> machine of the world, which was created by the best and
> most orderly of all artisans, had been established by the
> philosophers.

Any "lack of certitude" or irregularity in a system designed by "the best of all artisans" was deeply disturbing to Copernicus. He stated that "the mind shudders" at such a thing.

Furthermore, there were practical problems with the Ptolemaic system. While it was fairly accurate under most circumstances, it wasn't 100% accurate, and over the centuries since Ptolemy's death, enough astronomical observations had been made that didn't match with the Ptolemaic system that a growing number of astronomers were convinced something must be fundamentally wrong with the system. From a cultural standpoint, the most important area where the Ptolemaic system didn't exactly match observable reality had to do with the calendar. Julius Caesar (100–44 BCE) had decreed the year to be 365 ¼ days in length, but in reality it is a little over eleven minutes longer than that. While that doesn't sound like a lot, it meant that by Copernicus' time the calendar was off by ten days. That meant, among other things, that the Spring Equinox (the day on which every point on Earth has twelve hours of daylight and twelve hours of darkness) occurred on March 11th instead of March 21st, which is when it was supposed to occur. Since Easter is the first Sunday following the first full moon after the Spring equinox, that meant Easter wasn't being celebrated properly, which was deeply troubling not only to the papacy but also to other Christians knowledgeable about the problem. Copernicus was quite aware of this problem, and put in considerable time and effort attempting to resolve the problem while he was personal physician to his uncle, the bishop from 1503 to 1512 (he also received the post of canon in the church). Even though he never finished a medical degree at the University of Bologna, Copernicus would spend a great deal of time practicing medicine. While this job did include various duties not normally expected of a physician, such as occasionally acting as an ambassador, he also had considerable time to devote to astronomical research and to thinking about the problems of the calendar. His free time only expanded after the death of his uncle in 1512, when Copernicus moved to a sleepy little monastery in Frombork on the Baltic Sea coast.

At Frombork, Copernicus acted as a physician to the monks while carrying out dozens of astronomical observations and working his way through the problem of the calendar. It's likely that it was at Frombork where he wrote a short work known as the *Commentariolus* (Short Commentary), though it's also probable that he had developed the ideas contained in this work well before coming to the Baltic coast. This work lacks mathematical elaboration, but contains the essential insight that would drive Copernicus' work for the rest of his life: the idea that the sun, rather than the Earth, is the center of the universe (we would now say "solar system," but that isn't how Copernicus thought). How did Copernicus arrive at such a radical break with the Ptolemaic system? In all likelihood, it had nothing to do with celestial observation (since no observation available to him could have supported a heliocentric system) or mathematical reasoning. Instead, it was the result of his commitment to Hermetic philosophy.

In the Hermetic articulation of Neoplatonism, the One, which is said to have "exuded" the entire cosmos, is normally identified as the sun. This is as true in artistic representations of Hermeticism as it is of their writings. Christian Neoplatonists (and Hermeticism is an elaboration of Neoplatonism) had always identified the One as God, so Christian Hermeticists such as Copernicus identified the One with God, furthering identifying both with the sun. They didn't *literally* think the sun was God, but it did open the door to thinking about the sun as the center of all things, just as any good Christian thinks of everything as centered around God. It seems that with Copernicus, he took this idea more literally than many other Christian Hermeticists. But if God is the One, and the One is identified with the sun, why shouldn't the sun be the center of creation in the same way that God is? That's exactly what Copernicus suggested in his *Commentariolus,* which circulated among astronomers in manuscript form. We don't have to speculate where this notion came from; later Copernicus tells us in the first chapter of his *On the Revolutions of the Heavenly Spheres* that

> in the middle of all sits the Sun enthroned. In this most
> beautiful temple could we place this luminary in any better
> position from which he can illuminate the whole at once?
> He is rightly called the Lamp, the Mind, the Ruler of the
> Universe; Hermes Trismegistus names him the Visible God,
> Sophocles' Electra calls him the All-seeing. So the Sun
> sits as upon a royal throne ruling his children the planets
> which circle around him.

Few (if any) were convinced by the *Commentariolus*, but many thought it was a very interesting idea worthy of consideration, even though Copernicus' math didn't match observations any better than Ptolemy's did. Copernicus' system had the advantage of being somewhat simpler, though, and thus his math was easier to work out than Ptolemy's.

During this period, few knew about Copernicus' heliocentric beliefs. However, according to a rather curious passage from White, Copernicus (whom he refers to by the Polish form of his name, Kopernik), "had been a professor at Rome, but as this truth [the sun, not the Earth, is the center of the universe] grew within him, he seemed to feel that at Rome he was no longer safe." That is an odd thing to write, because while Copernicus spent 1500 in Rome, this was well before he finished his studies and he certainly wasn't there as a professor; at best, he had some sort of internship, and there's no indication that he felt any discomfort at being in Rome. White goes on to write that Copernicus allowed his ideas to go unpublished for more than thirty years because he dared "not send [his book] to Rome, for there are the rulers of the older Church ready to seize it," and in one of his rare references to Protestantism, he goes on to write that Copernicus dared "not send it to Wittenberg, because there are the leaders of Protestantism no less hostile." Draper has little to say about Copernicus, dealing with him in only three paragraphs, but he indicates agreement with White, writing

> Aware that his doctrines were totally opposed to revealed truth, and foreseeing that they would bring upon him the punishments of the Church, he expressed himself in a cautious and apologetic manner, saying that he had only taken the liberty of trying whether, on the supposition of the earth's motion, it was possible to find better explanations than the ancient ones of the revolutions of the celestial orbs; that in doing this he had only taken the privilege that had been allowed to others, of feigning what hypothesis they chose.

Many modern scientists such as Neil deGrasse Tyson are as sure as White was that the Catholic Church was to blame for making Copernicus fearful and hesitant. In the first episode of *Cosmos: A Spacetime Odyssey,* "Standing up in the Milky Way," Tyson makes only a passing reference to Copernicus as a man "who didn't go far enough." Sagan had only a little more to say about Copernicus, stating that he "annoyed a lot of people" and noting that the Catholic Church placed his book *On the Revolutions* "on its list of forbidden books," without mentioning that this event didn't happen for more than a century. One gets the idea these two astrophysicists see Copernicus as something of a failure, who was too frightened of the Catholic Church to push his ideas to their natural conclusions. That isn't a very accurate view of the historical Copernicus, though.

So what's the truth about Copernicus? First of all, his idea about a heliocentric system was widely known during his lifetime, at least among intellectual circles. Pope Sixtus IV (r. 1471–1484) consulted Regiomontanus about the problems of the calendar in 1475, and when he offered no solution, in 1514, Pope Julius II asked a number of scholars and astronomers how to resolve the problem, including Copernicus. The papal request was based on the reputation Copernicus had gained through the letters he wrote to astronomers across the continent as well as the ideas contained in the *Commentariolus.* Thus, Pope Julius II was well aware of Copernicus' attractions to heliocentrism and seems to have been intrigued, rather than concerned. Copernicus declined to get involved with the project, stating that until the motions of the sun and moon were better understood, it would be impossible to rectify the calendar. Copernicus continued to tinker with his system over the following decades, but never solved the problems of planetary motion, largely because he persisted in thinking of orbits as being perfectly circular.

In spite of Copernicus' failure to perfect his system, there were astronomers who took notice. In 1538 an Austrian mathematician, astronomer, and astrologer named George Joachim Rheticus (1514–1574) left the University of Wittenberg where he had been studying, possibly because of a bit of trouble he'd gotten into for his association with a radical poet who had been expelled. Rheticus took this opportunity to travel to Nuremberg, one of the most important centers for publishing in Europe. It was there where he learned about Copernicus' ideas and decided to travel to Frombork to learn more. Although Rheticus was a Lutheran and Copernicus a Catholic and a church canon, in these early days of the Reformation scholars paid little attention to such differences. Rheticus took bound copies of some of the most cutting-edge astronomical observations as a gift for the older man. Copernicus seems to have been flattered, and Rheticus studied with him until 1541, being the only student Copernicus ever had. Rheticus urged Copernicus to publish, or let him publish his ideas, but Copernicus refused—not because of fear of getting into trouble with the church, but because he wanted to incorporate the new information Rheticus had brought him to improve the accuracy of his system, but could never quite get the level of accuracy he desired. Copernicus did agree, however, to allow Rheticus to publish a short report, the result of which was the *Narratio prima*, or *First Report*. In this work, Rheticus referred to Copernicus as an astronomer equal to Ptolemy for developing a cosmological system that didn't require the equant. Most importantly, according to Rheticus, Copernicus' heliocentric system was "internally harmonious" and completely rational, claims that had considerable impact on humanist scholars across Europe.

Copernicus was very slow to publish, though. In 1541, a Lutheran minister by the name of Andreas Osiander (1498–1552) wrote to Copernicus. He had studied at Wittenberg University, where Rheticus had studied, which would become an early center of interest in Copernican astronomy. Osiander wasn't a professional astronomer, but he had a talent for mathematics and had studied astronomy at university. Osiander suggested Copernicus include a few words in his work indicating that his heliocentric theory was merely a mathematical model and not intended to represent a real model of the universe, so that he might "mollify the peripatetics and theologians whose opposition you fear." This is the only reference to any sort of concern Copernicus might have felt about publishing. The term "peripatetics" is an old one, referring to Aristotelian philosophers, and meant to call to mind traditional-minded university professors. Thus, while Osiander does indicate a concern about how theologians might view Copernicus's book—and he suggests that Copernicus shared the concern—he first mentions the opposition the book might face from academics. We should note, however, that we have no evidence that Copernicus himself ever expressed concern about how theologians might view his book

Whatever reservations Copernicus might have felt, Rheticus finally convinced him to publish, in light of the generally positive reception received by the *First Report*. Aging and ailing, Copernicus likely realized he didn't have much longer to get his ideas published. Rheticus took a leave of absence from the University of Wittenberg to edit Copernicus' work, double-check his math, and get it to a printer. The book finally appeared in print in 1543 as *On the Revolutions of the Heavenly Spheres*, dedicated to Pope Paul III (1534–1549). Copernicus's concerns seem to have been poorly founded, as the church expressed no negative reaction, at least at first, and the scholars who reacted negatively were few and far between. As we'll see in the next chapter, the negative reaction that occurred later was not entirely based on the content of Copernicus' work.

However, there was another problem with *On the Revolution*. After doing a great deal of work on the book, Rheticus had to leave Nuremberg in order to take up a position at the University of Leipzig, so he left Osiander in charge of doing the final proofreading. Osiander appears to have unilaterally decided to write a preface, in which he undercut the central premise in *On the Revolutions*. Osiander wrote that the heliocentric system was presented merely as a mathematical device and wasn't intended to indicate that the Earth really moves. Anyone who read the work and failed to understand that point, would "depart from the study a greater fool than when he entered it." The worst thing about this preface was that it was included without attribution to Osiander. Therefore, many people thought Copernicus himself had written it. Osiander wasn't intentionally sabotaging Copernicus' work; he seems to have had the best intentions. He was worried about

opposition from conservative theologians who might see a cosmological model in which the Earth moved as a violation of scripture, but he was also acutely aware that astronomers might take
exception to a system that flew in the face of so many centuries
of European scholarship. A fundamental premise of Aristotelian
physics is that things fall because they possess the element
"earth" and are therefore drawn to the center of the Earth. A
moving Earth would negate all Aristotelian physical principles.
And his preface did have the effect of making Copernicanism
more palatable to scholars; though over the course of the next
century few actually accepted Copernicus' model representative
of the way the universe is actually constructed, many scholars
used Copernicus' math when it proved easier than Ptolemy's and
praised the Polish scholar for providing a means of getting rid of
the equant. However, most scholars continued to accept some
version of the geocentric system as the way the universe *actually*
worked. While that may strike modern people as odd, the reason
why is because we grew up being taught that Copernicus was
right. There was no observational data available in the fifteenth
or sixteenth century to support the heliocentric system, and while
his math was easier to work out, it's predictions weren't more
accurate than those provided by the Ptolemaic system.

As for theologians, the response was, at least at first, a deaf-
ening silence. Among Catholics, Pope Paul III (r. 1534–1549), to
whom *On the Revolutions* was dedicated, ignored Copernicus.
His trusted advisor, the Dominican Professor of Theology at the
University of Bologna, Bartolomeo Spina of Pisa (1474–1546), may
have been planning a denunciation of Copernicus before he died,
but if so, no one else took Spina's work up. As for the Protestants,
one of Martin Luther's (1483–1546) students reported that Luther
stated: "[t]hat fool [Copernicus] would upset the whole art of
astronomy." However, that report is from the *Table Talk,* published two decades after Luther's death. It's not at all clear this
statement is an accurate representation of Luther's views, or even
if it meant that he disagreed with Copernicus. He may have simply been stating the fact that his heliocentric model was going to
be disruptive. Certainly, his right-hand man Philip Melancthon

(1497–1560), who was responsible for the redesign of the curricula at the universities of Tubingen, Wittenberg, and Leipzig,
held no reservations about using Copernicus' math. Thanks to
Melancthon's curriculum redesign, Wittenberg became the primary center for the study of Copernican astronomy in Europe.
In short, by 1600 there was no official theological position from
either Protestants or Catholics regarding Copernicus, and no real
reason to think the theologians had a problem with his heliocentric system.

Beyond Rheticus and Maestelin, who accepted heliocentricism
on mathematical grounds, there was at least one who accepted it
for entirely different reasons: Giordano Bruno (1548–1600). We
ran into him in the introduction, when we pointed to Neil deGrasse
Tyson's 2014 version of *Cosmos*. In the very first episode, Tyson
intoned that Bruno was the only man on "the whole planet who
envisioned an infinitely grander cosmos" than his contemporaries,
a man whom the Catholic Church imprisoned and eventually executed because he "couldn't keep his soaring vision of the cosmos
to himself" during a time when "there was no freedom of thought."
It's not hard to figure out where Tyson (or the show's writers) are
getting this image of Bruno from, especially since Tyson actually
cites White's *A History of the Warfare of Science with Theology
in Christendom* in his own essay, "Holy Wars: An Astrophysicist
Ponders the God Question." And as a reminder, Tyson is rather
uncritical in the way he uses this source, stating that it "reveals
a long and combative relationship between religion and science."
Since there are a large number of far more recent historical studies dealing with this topic—I can't recommend Ronald L. Numbers'
collection of essays on *Galileo Goes to Jail and Other Myths about
Science and Religion* strongly enough—it's not clear why he didn't
grab one of those books. Many of them are more readable and
most of them do a better job with the history.

However, for Tyson's purposes, there is one thing White does
better than more recent and knowledgeable scholars such as
Numbers—he presents a history that is in line with what Tyson,
as a modern scientist, wants to believe. In White's soaring vision
Bruno was a hero, and Copernicanism was,

the new truth [that] could not be concealed; it could neither be laughed down nor frowned down. Many minds had received it, but within the hearing of the papacy only one tongue appears to have dared to utter it clearly. This new warrior was that strange mortal, Giordano Bruno. He was hunted from land to land, until at last he turned on his pursuers with fearful invectives. For this he was entrapped at Venice, imprisoned during six years in the dungeons of the Inquisition at Rome, then burned alive, and his ashes scattered to the winds. Still, the truth lived on.

Wow. That's certainly a stirring and heroic image, and one that many scientists now fully accept, and what's not to love? We all need heroes, after all. I was certainly moved when I first read about Bruno in Isaac Asimov's *Chronology of Science and Discovery.* Many know Asimov, who died at the age of seventy-two, as a science fiction writer, but he held a Ph.D. in organic chemistry and wrote on almost everything (he is listed in the *Guinness Book of World Records* as the most published author in history), and his presentation of Bruno as a martyr for science was one I accepted for many years. However, a funny thing happened during my years of graduate school and beyond—I actually looked into the history of Giordano Bruno and read many of the primary sources associated with him. In so doing, I met a very different man than I expected to encounter, based on what I'd read in Asimov's *Chronology,* and a wholly different man than Tyson presents in *Cosmos.* So who was the historical Giordano Bruno?

First of all, he didn't become "Giordano" until he was ordained as a priest. He was born Fillipo Bruno in the Kingdom of Naples in southern Italy. The son of a soldier, for unknown reasons he was educated at the Augustinian monastery in Naples, which likely meant that he impressed his local priest as a bright boy who deserved more education than his soldier father could provide. He also attended public lectures at the Dominican *studium generale* in Naples. At seventeen he donned the habit of the Dominicans, and at the age of twenty-four he was ordained a priest. However, even at this early stage in his life he held unconventional beliefs, and got himself into trouble for ridding his cell of images of the saints and keeping only a crucifix, as well as recommending questionable works to other novices. What really got him into trouble, though, was reportedly defending the Arian heresy—which rejected the Trinity—and for keeping an annotated copy of Erasmus' (1466–1536) critique of the Catholic Church hidden in a toilet. He learned that he could soon expect a visit from the local inquisitor, so he temporarily tossed his Dominican habit aside and fled Naples for northern Italy, where he took his habit up again for a time, and then across the Alps into Geneva, which was now a staunchly Protestant city. He may have briefly converted to Calvinism while there, but he definitely enrolled as a student for a brief time at the University of Geneva.

As a student in Geneva, Bruno demonstrated his one true gift apart from an apparently prodigious memory—the ability to offend important people. The chair of the department of theology at Geneva was a man named Antoine de la Faye (1540–1615), whose ethics and teaching competence were both open to question. What seems to have most irritated Bruno, however, was that La Faye taught entirely through the use of lecture and wouldn't engage his students in debate. Therefore, Bruno wrote up a pamphlet of twenty errors he said La Faye had made in a single lecture and had it published, getting both himself and the publisher into considerable hot water with the politically well-connected La Faye. On August 6th, 1579, the Consistory (the governing body of Geneva) had both Bruno and the printer thrown into jail. After almost three weeks in jail, Bruno confessed to slander, agreed to a fine and to accept excommunication from Geneva, and quickly fled to France. This was but the first of many such times he would be forced to flee regions where he angered influential men. For the next few years, he lived in France and managed to obtain a Doctorate in Theology from the University of Toulouse. He also taught in Lyons, Toulouse, and Paris, as well as published three short works on memory, as well as a comedy.

Apparently tired of living a quietly successful life, Bruno left France for England in 1583 where he stayed for two years, which was roughly long enough to anger more or less every intellectual in the country. Although he was ridiculed by George Abbot (1563–1633) for maintaining "the opinion of Copernicus that the earth did go round, and the heavens did stand still; whereas in truth it was his own head which rather did run round, and his brains did not stand still," it was his attacks on Aristotelian natural philosophy and his caustic personality that got him into trouble. Somewhere along the way, Bruno had become an adherent of Hermetic philosophy, and as such, rejected Aristotle. That was fine as long as he was on the Continent, where Plato had come to either dominate or to be held in high regard, depending on what part of Europe one was talking about, but at English universities Aristotle still reigned supreme. Furthermore, Bruno had become deeply interested in Hermetic magic, which is why he was accused of plagiarizing Marsilio Ficino while in England. Ficino, in addition to being a translator of Plato, wrote the *Three Books on Life*, which was replete with magical ideas based on the doctrine of affinities—the idea that certain things in nature have an unseen but real connection to certain other things, and thus by carrying out a change in one thing, like a diamond, a change in another thing, like a kidney stone, might be brought about. While certain usages of this form of magic, like the example of a medical use I just gave, weren't seen as controversial, other uses were frowned upon or viewed with outright fear. For someone such as Bruno who collected enemies the way some people collect coins, dabbling with Hermetic magic was quite dangerous.

After a rather dramatic incident in which a mob stormed the French embassy in London, Bruno fled back to France in 1585, where he again ran into trouble for attacking Aristotelian natural philosophy, and more importantly for attacking the Italian mathematician, Fabrizio Mordente (1532–1608) in a work titled *Idiot Triumphant*. The two exchanged literary barbs, but unfortunately for Bruno, Mordente's patron was the politically powerful Charles, Duke of Guise, who was the son of the French king, and Bruno was forced to flee back to Italy in 1592. He taught briefly at Padua, and applied for a position as a mathematics professor at the university, a position that went to Galileo (1564–1642), whom we'll discuss in the next chapter (and who incidentally held a very low opinion of Bruno), before going on to Venice at the invitation of the wealthy and politically powerful, Giovanni Mocenigo (c. 1558–c. 1623). Mocenigo hired Bruno to teach him his memory tricks, but apparently was deeply dissatisfied with his teaching and denounced him to the local inquisitor.

What happened next is admittedly not completely clear, as the records regarding the investigation of Bruno and his subsequent trial are incomplete. That isn't the result of some nefarious plot on the part of his inquisitors, however, but rather the result of the natural effect of centuries of history. Try to remember where your tax records from ten years ago are for just a moment. The odds are good that you have no idea—such things become lost. Now imagine that the records we're talking about are more than 400 years old and you'll see the problem. What records we do have are largely thanks to the Catholic Church; in 1940 the Archivist of the Vatican Secret Archives and Librarian of the Vatican Library, the Cardinal Giovanni Mercati (1886–1957), found a number of records relating to Bruno, including a summary of the proceedings against him. One thing we should keep in mind, however, is that Bruno was held for seven years, during which time he would have been given many opportunities to recant—that is, admit he was wrong and take back his heretical statements. In spite of what modern people might imagine about inquisitors, they didn't normally want to convict anyone of heresy. They far and away preferred to teach heretics the error of their ways and have them recant, and that's what happened in the majority of cases where an inquisitor decided a case of heresy was considered proven. Thus, while the actions of inquisitors may seem horrible to modern people—and as a strong proponent of free thought, I fully understand that feeling—we should judge their actions in light of the time and place in which those actions occurred, and we should try to understand what they were trying to accomplish— which was decidedly *not* to execute heretics, in the vast majority of cases. The fact that the investigation of Bruno lasted for seven years speaks to a desire to be thorough and fair.

However, in the end Bruno refused to recant and was condemned to death. Among the charges he was convicted of, we do find that he maintained the "plurality of worlds." In other words, he speculated that each star was a sun around which worlds with intelligent inhabitants might be found. However, he was far from the first to put such speculations forward—the fourteenth-century theologian Nicole Oresme and the fifteenth-century Cardinal Nicholas of Cusa also advanced such speculation. What made Bruno different was that he went on to suggest that the cosmos was truly eternal and uncreated, which was a problem for the church. However, this charge was far down the list of those brought against Bruno. Far more importantly, he also argued that Jesus wasn't the son of God, but was instead simply a human magician, that Mary had never given birth as a virgin because her son wasn't the son of God, that the bread and the wine didn't transform into the body and blood of Christ during mass, that upon the death of humans their souls might be reborn into new bodies, even the bodies of animals, and finally, last but not least, that he was involved in forbidden forms of magic. That last charge makes it likely that he was believed to be a necromancer, a form of magic that by the thirteenth century was seen as involving attempts to summon demons or otherwise gain the help of demons. Since any such act was believed to require a pact with Satan, by definition from the end of the fourteenth century on this form of magic was seen as a form of heresy, because it required the magician to reject God and swear allegiance to Satan.

The charge of being involved in forbidden forms of magic may in fact hold the key to why Bruno was prosecuted and why he refused to recant. As the work of Richard Kieckhefer, Michael Bailey, and Benedek Lang all make abundantly clear, there was a clerical underground of priests who believed in and attempted to use black magic for their own benefit from at least the thirteenth century on. Therefore, while there's no concrete proof that Bruno was involved in such work, it isn't beyond the realm of possibilities and would explain why he wouldn't simply confess and recant. Why would he recant if he didn't believe in the power of God? While such speculation can't be proven, we're on much firmer ground in saying that Bruno most emphatically and decidedly was *not* prosecuted for advocating Copernican beliefs, because such beliefs were in no way forbidden in 1600. Nor was he prosecuted merely for stating that there might be many worlds, or even for suggesting that the universe is infinite. The English natural philosopher Thomas Digges (1546–1595) had first put that idea into print in 1576, and though he was a Protestant, Catholic astronomers on the Continent were aware of and discussed his ideas without incurring any penalty because it simply wasn't a particularly controversial position.

Furthermore, it's equally easy to establish that Bruno was no scientist. When he wrote about Copernicanism, for example, he didn't bother with providing a mathematical basis for his belief in heliocentrism. That wasn't because mathematics was beyond him. After all, he had studied the same seven liberal arts all other university-educated men studied, which included mathematics, astronomy, and geometry, and at times he wrote a fair bit about magic. And when he wrote about magic, such as in his *De monade numero et figura* (*On the Number and Figure of the Monad*) he would always include the mathematics he felt provided a necessary basis for magic. However, neither mathematics nor empiricism—and empiricism is something Bruno would have rejected out of hand, as he was a firm believer that real truth was to be found in a metaphysical world beyond the mundane physical one in which we exist, as we imagine ourselves to exists, as Bruno might have put it—had anything to do with his heliocentrism. He was a heliocentrist for the same reason that Copernicus was, because it fit better with his Hermetic beliefs. Hermeticism, remember, identified the One, the source of all existence, with the Sun, which made a heliocentric system seem logical to Hermeticists such as Bruno and Copernicus. Bruno doesn't even seem to have understood Copernicus' system all that well, as evidenced by mistakes he made when discussing it. Frankly, Copernicus' work was probably too technical for Bruno to understand. Copernicus actually developed the mathematical models to support his system, whereas Bruno just wrote about heliocentrism while slinging insults at anyone who disagreed with him, didn't show him the level of respect he felt he was due, held positions he didn't support, or any of a number of other reasons.

In the end, on February 17th, 1600 the secular authorities—to whom the papal inquisition had turned Bruno over—burned Giordano Bruno at the stake in the Campo de' Fiori, a central market square in Rome. Bruno was a difficult man throughout his life who made enemies of many powerful men, earned few friends, and may have dabbled in black magic. He definitely rejected central tenets of the Catholic faith, including the divinity of Christ, Mary's virginity, and the transubstantiation of bread and wine in the mass. Given how touchy the Church had become over such things in the wake of the Protestant Reformation that began with Martin Luther nailing his 95 Theses to the church door in Wittenberg (by 1600 wars of religion had killed millions in both Germany and France), one need look no further than these circumstances to understand why Bruno was executed.

The Renaissance was a time when intellectuals and artists basked in the patronage and respect of many people. That respect translated into a growing recognition of the value of manual activities such as painting and sculpting, which would later translate into an idea that gentlemen could indeed engage in work with their own hands—as long as that work had cultural merit. We'll see that change come to fruition in the Scientific Revolution, the subject of the next chapter. Furthermore, cultural changes that included competition between city-states and nations and the rise of an increasingly wealthy group of bankers and merchants who longed for legitimacy and respect and were willing to pay for it meant that scholars, including natural philosophers, had many potential sources of support. That included the Catholic Church, which supported, rather than oppressed, the work of scholars such as Copernicus.

Further Reading

Bailey, Michael D. *Battling Demons: Witchcraft, Heresy, and Reform in the Late Middle Ages.* University Park: The Pennsylvania State University Press, 2003.

Faivre, Antoine and Wouter J. Hanegraaff. *Western Esotericism and the Science of Religion.* Leuven: Peeters, 1998.

Gatti, Hillary. *Giordano Bruno and Renaissance Science.* Ithaca: Cornell University Press, 1999.

———. *Essays on Giordano Bruno.* Princeton: Princeton University Press, 2011.

Kieckhefer, Richard. *Forbidden Rites: A Necromancer's Manual of the Fifteenth Century.* University Park: Pennsylvania State University Press, 1998.

Kuhn, Thomas S. *The Copernican Revolution: Planetary Astronomy in the Development of Western Thought.* Cambridge: Harvard University Press, 2003, 24th printing.

Lang, Benedek. *Unlocked Books: Manuscripts of Learned Magic in the Medieval Libraries of Central Europe.* University Park: Pennsylvania State University Press, 2008.

Neuber, Wolfgang, Thomas Rahn, and Claus Zittel. *The Making of Copernicus: Early Modern Transformations of a Scientist and his Science.* Leiden: Brill, 2014.

Olson, Richard G. *Science and Religion, 1450–1900: From Copernicus to Darwin.* Baltimore: Johns Hopkins University Press, 2006.

Peters, Edward. *Inquisition.* Berkeley: University of California Press, 1989.

Yates, Frances. *Giordano Bruno and the Hermetic Tradition.* Abdingdon: Routledge, 2001, 3rd printing.

CHAPTER 4

The Scientific Revolution in Germany and Italy

T he Scientific Revolution as an idea has taken a bit of a beating among historians over the years. As I noted in the last chapter, there's no way to pin it down to one period. Copernicus was definitely a product of, and lived during, the highpoint of the Renaissance and overlapped with the beginning of the Protestant Reformation. Similarly, the primary protagonists of the Scientific Revolution lived during the immediate aftermath of that Reformation and the Catholic Reformation that occurred as a response, and frankly their lives can't be understood without acknowledgement of that background. More troubling for the idea of a revolution, is that it occurred over a very long period of time. Almost 200 years lie between the lives of Copernicus and Isaac Newton (1643–1727), and the changes that occurred happened slowly, often not the way they are related in standard history books, and frequently for reasons that don't match up to the perception of the Scientific Revolution as a triumph of rationalism and empiricism over mysticism, which is the way Carl Sagan would have it; nor was it quite built on the "the refusal to believe any explanation of natural phenomena that could not be proven to the satisfaction of the empirical observer" as the surgeon and science writer Sherwin Nuland (1930–2014) described. In fact, many of the scientific advances that occurred during the Scientific Revolution occurred because of a commitment to religious belief (or mysticism, as Sagan would have put it) and were accepted *because* of those beliefs, rather than Nuland's strict empiricism. And though it was strung out over a long period of time and followed many odd twists and turns, if Copernicus could have time traveled forward to observed Newton's funeral in 1727, he wouldn't have recognized the ways in which scholars studied the natural world, developed proofs, or conceived of the natural world. Things had changed so thoroughly that it's impossible to call this period anything other than revolutionary. However, because the events were strung out over such a long period, and are so central to the story I'm telling, in this chapter I'll deal only with the early part of the Scientific Revolution in German-speaking lands before moving on to Italy, where I'll focus considerable attention on Galileo, because of the importance of his story to understanding the relationship between religion and science in the West.

This was an era of turmoil; Germany had been ground zero for papal fund-raising activities for centuries to the point that one fifteenth-century German priest had referred to the region as "the Pope's cow," because every time a little money was needed, the papacy milked the Germans for it. This was because Germany, unlike France, England, and Spain—which all had highly centralized governments that placed sharp restrictions on papal fundraising because the rulers of these kingdoms didn't want competition for their own efforts to milk their populations of money—was a decentralized mess. As the center of the Holy Roman Empire, Germany was made up of over 200 different semi-independent regions, loosely held together in what was in actual fact a confederacy rather than an empire, governed by an elected emperor who was only as powerful as his personal qualities would allow. He was never powerful enough to keep the pope's indulgence sellers from traveling back and forth across the land, convincing the inhabitants to buy tokens that would presumably reduce their time in purgatory (or the time of a dead family member) by fair means or foul.

The activities of these indulgence sellers were always on shaky theological footing, which only became worse as the papacy pressured them to bring in more cash in the sixteenth century. A series of disasters had affected the prestige and income of the papacy, including a period when the pope was in Avignon rather than Rome, followed by a period where there was first two popes, then for a brief time, three. In order to rebuild that prestige, sixteenth-century popes used all the Renaissance tricks described in the last chapter, from supporting scholars and artists to indulging in massive building projects—none larger or more expensive than razing the tomb of St. Peter and building a bigger, grander structure, St. Peter's Basilica, begun in 1506. As indulgence sellers put their activities into high gear, a German professor of theology inadvertently kicked off the Reformation by issuing a call for reform of the church in 1517 that had morphed into an outright break with Rome by 1521. A generation of religious war followed, with bloody, though intermittent, conflict occurring from 1525 to 1555 in Germany. During this period a dizzying array of Protestant leaders emerged, from Jean Calvin (spiritual father of a variety of groups all loosely called Calvinists, from the Huguenots in France to the Presbyterians in Scotland, 1509–1564) in France to Huldrych Zwingli (founder of what would become the Baptist movement, 1484–1531) in Switzerland.

The religious reformers were linked to the Renaissance by taking an *ad fontes* approach to the bible, reading it for themselves in the original Greek of the New Testament and the Hebrew of the old, which led to interpretations of its meaning that varied widely. These differing religious interpretations sparked or contributed to more wars, in France (1562–1598), various territories of the Holy Roman Empire, and England, where a civil war that began in 1642 wasn't strictly a war of religion but was made more complex by religious divisions in the country. Other than the Renaissance interest in languages and close reading of texts, Renaissance art and scholarship spread northward and intermingled with the various forms of Christianity that now existed across Europe, and the resulting Christian humanism had many cultural effects, from an increase in self-confidence of these humanists to a slow but steady growth in recognition that even people of good birth could engage in manual activities without compromising their status, which resulted from the growing social and cultural importance of artists.

From a cultural standpoint, we should keep in mind that the Catholic Church in 1500 was relatively complacent, in many ways. The authorities tolerated divergent beliefs more than modern people normally think was the case, within limits—denying the divinity of Christ, the virginity of Mary, or the Trinity could get one into trouble with inquisitors rather rapidly. From time to time people produced translations of the sections of the bible into vernacular languages, and as long as this was done by a priest who made a good-faith effort to be true to the text, no one was very bothered by such activities, though there wasn't a great call for translations because literacy was slow to develop among the laity. Furthermore, the medieval and Renaissance church stressed allegorical reading of the bible over literalist interpretations. However, with the advent of the Reformation Catholic, authorities became much more nervous about beliefs or practices that didn't square with those they approved of, and translations of the bible in whole or in part began to arouse considerable suspicion—not least because many translators added commentary in the margins with a frankly anti-Catholic stance, and often followed translation practices that resulted in anti-Catholic texts. Perhaps most importantly, at the Council of Trent (1545–1563), which was a meeting of hundreds of theologians and leaders of the church who met in northern Italy in order to create policies for how to deal with the Protestants while reforming the church, church leaders decided that maybe the Protestants were onto something about the bible, and announced that the text of the bible should be taken literally as often as possible. This was quite a new stance for the Catholic Church, and would prove to be troublesome.

The period of history we're going to cover in this chapter is both vast and wide ranging, giving us a million possible starting points. But because every story must start somewhere, we'll begin ours in the sleepy little town of Weil der Stadt in what is now southern Germany in the late sixteenth century. Our focus will be the grandson of the town's Lord Mayor and son of a mercenary soldier who most likely died when fighting in the 80 Year's War (a struggle between the Calvinist Dutch and their Catholic Spanish overlords, which was made more complex and deadly by the religious divisions, even if it wasn't completely driven by them) who worked alongside his mother at his grandfather's inn, occasionally astounding patrons with his mathematical abilities. Johannes Kepler (1571–1630) was a sickly child, with eyes and hands that had been weakened by the smallpox that had almost killed him as an infant, but with a mind that was quick and ever-questioning. His mother fostered his interests, which came to focus on astronomy from an early age, by taking him "to a high place" to see the comet of 1577 when he was six, as he later wrote, and calling him outdoors to see the lunar eclipse of 1580.

However, because of his childhood bout with smallpox, Kepler's eyes were far too week to make the naked-eye observations necessary to work as an astronomer in this period before the invention of the telescope. That, combined with his intense religiosity, may have been why Kepler dreamed from an early age of becoming a Lutheran minister. Thanks again to his supportive mother, as well as his natural intelligence and the vestiges of the family's connections, Kepler was able to transfer to the Lutheran seminary at nearby Maulbronn once he had finished grammar school when he was fifteen. This former Cistercian monastery had been converted into a school to train future Lutheran ministers in 1556, and though the aim was religious, the curriculum was organized along the lines Luther's right-hand man and guiding intellectual light Philip Melancthon prescribed. Thus, Kepler got a solid grounding in mathematics and astronomy, which likely included instruction in both the new Copernican system as well as traditional Ptolemaic astronomy. Whether or not he learned about heliocentrism at Maulbronn, he certainly did when he entered the University of Tubingen at seventeen.

Still intent on becoming a minister, Kepler studied theology under a student of Melancthon's, but he also learned a great number of other things. Of particular importance was Hermetic philosophy, which would have a huge influence on his life. One of his professors was Vitus Müller (1561–1626), who judging by his surviving writings was primarily an Aristotelian with a particular interest in moral philosophy and ethics. However, in keeping with the times, he was as fluent in Greek as he was Latin and insisted on an *ad fontes* approach to the study of the classics, and it's possible that it was through him that Kepler read the *Corpus Hermeticum*. Regardless of who introduced him to the so-called Hermes, though, it's obvious that he read these works closely and with great interest, as he would quote Ficino's translation of the *Corpus* from time to time in the years to come. He also made a strong association between the One of Hermetic Neoplatonism, that central entity from which all existence flows, and the Trinity, in a very similar fashion to Copernicus. Finally, he also became entranced with the Hermetic idea of the "Cosmic Harmony," the notion that the rational, mathematically precise way in which God had ordered the heavens meant that they moved in a harmonic relationship with one another, in precisely the same fashion in which music is ordered. Thus, the motions of the heavenly bodies produce a non-audible, but nevertheless very real form of music, the importance of which is that there are mathematical relationships identical to the types of relationships that exists between musical notes, waiting to be discovered by the person with the right combination of mathematical skill and philosophical/religious insight.

Kepler's religious sensibilities drew him to Hermetic philosophy, which in turn drew him to astronomy. His instructor was Michael Maestlin, who taught at Tubingen for an impressive forty-seven years, from 1583 until just before his death in 1631. We met Maestlin in the last chapter, because though he wrote a popular textbook on Ptolemaic astronomy, he was the first professional astronomer since Rheticus to accept heliocentrism as a realistic representation of the cosmos, instead of simply a mathematical model. Just as Copernicus had been attracted to such a system by his interest in Hermeticism, so was Kepler. As Kepler would state in 1593, during a formal disputation (discussion-based arguments were central to university education) with fellow physics students, "The sun . . . appears, by virtue of his dignity and power, suited for this motive duty (of moving the planets) and worthy to become the home of God himself." And just as Copernicus had been convinced of the orderly and rational structure of the universe because of his belief in an orderly and rational God, so was Kepler. Granted, Kepler was a committed Lutheran whereas Copernicus was a Catholic, but these confessional distinctions had no bearing on how they viewed God.

However, as a student Kepler was almost wholly fixated on his professional goal of becoming a Lutheran minister. Regarding astronomy and mathematics, he later wrote that they were simply part of the "prescribed studies and nothing indicated to me a particular bent for astronomy." He completed his Master of Arts degree and entered into the three years of study required to become a minister, but in his last year an event happened that would prove to be life changing for Kepler—the mathematics instructor at the Lutheran school of Graz in modern-day Austria died, and the school administration asked Tubingen to provide a replacement. Though he complained bitterly about being ordered to give up his theological studies, the university administration chose him to take the post and he did so at the age of twenty-two.

To be a good teacher requires a combination of personal qualities, training, and education. It helps to be outgoing, though introverts can learn to be good teachers, and of course, one needs to be knowledgeable about the subject being taught. Even more important than simply being knowledgeable, a teacher must also be confident in his her or knowledge. I've personally known too many brilliant and knowledgeable people whose lack of confidence sabotages them as teachers. But one of the most important elements if one is to truly sparkle in the classroom is to have a love not only of learning, but also of teaching, and a joy in watching young minds develop and the patience necessary to help students develop the skills and knowledge that are already second nature for any competent teacher. Of all these qualities, Kepler had only knowledge. He was not outgoing, had not received training in how to overcome that—or how to teach at all, for that matter—and worst of all he had no real interest in teaching, nor did he have the patience required to teach the middle-school-aged children to whom he was assigned. However, Kepler also met a twenty-three-year-old widow with a young daughter named Barbara Müller, who "set his heart aflame," and he began to pursue her in 1595. So with no other realistic career options, he doggedly kept at his teaching.

However, Kepler's limited interest and skill in teaching left him with considerable intellectual energy that required an outlet. And since his love interest, Barbara Müller, had inherited a bit of land and money from the two husbands she had outlived, her father didn't feel Kepler was worthy of her, as he had little money of his own. Thus, he needed some way to distinguish himself, and an opportunity arose when he had an epiphany while lecturing on the periodic conjunction of Saturn and Jupiter in the zodiac to a no-doubt bewildered class of pre-teens. It suddenly occurred to him that there was a definite mathematical relationship between the inner boundaries of regular polygons and circles, and he reasoned that this might provide a geometrical basis for a heliocentric system. While he was unable to work out any such relationship, he hit upon the idea of using the five Platonic solids (so called because Plato had suggested that the classical elements might be made up of these solids), and remarkably the math worked. In the model that Kepler produced, each planetary orb (the crystalline sphere that represented the circular orbit of the planet) fit within one of the 5 three-dimensional polyhedra (the tetrahedron, cube, octahedron, dodecahedron, and icosahedron—which look, coincidentally like a set of dice for the game *Dungeons and Dragons*), placed at intervals that corresponded to the size of each planet's path around the sun, with each solid nesting within another like a set of Russian nesting dolls. This produced a model with remarkably precise predictive power for the motions of the planet. Even better, there are five Platonic solids, and when one allows for a sphere encasing all five, the number of Platonic solids fit perfectly with the fact that there are six planets observable with the naked eye: Mercury, Venus, Earth, Mars, Jupiter, and Saturn. Furthermore, the idea that each of the planets is embedded in a sphere nested within a polyhedral solid, which is itself nested inside another solid, provided the mechanism for planetary motion. It's the motion of the polyhedron that move the spheres, which moves the next polyhedron, and so forth. As far as Kepler was concerned, this geometrical proof and the mathematical precision for which it allowed proved that God created the universe in His image, giving it the rationality that Kepler knew *must* be there.

With Maestlin's support, Kepler appealed to the University of Tubingen to publish his work, which it did, once he removed passages of biblical commentary the theology faculty at Tubingen felt Kepler unqualified to supply. Published in 1596, this book wasn't widely read but it did establish Kepler's reputation as an astronomer and demonstrated to Barbara Müller's father that Kepler was a worthwhile match for his daughter, causing him to allow them to marry in 1597. For the rest of his life Kepler never gave up on the notion of a direct correlation between the planets and the Platonic solids. Hopefully readers have already caught on that this apparent relationship is simply coincidental, but that's actually the point: Kepler's faith that God had created the universe according to a mathematically ordered rationality drove him to search for the keys to that order, and in some cases to see it where it didn't exist—but in many others, to hit upon important understandings about the nature of the universe that would, well, revolutionize astronomy. Kepler wrote Maestlin in 1595, "I wanted to become a theologian . . . for a long time I was restless. Now, however, behold how through my effort God is being celebrated in astronomy." Although many modern scientists feel that religious beliefs oppose scientific ways of thinking—as evidenced by Carl Sagan's statement about the times in which Kepler lived, made in the third episode of his version *Cosmos* titled "The Harmony of the Worlds," that the "human spirit [was] fettered and the mind chained" by religion—that was certainly not true for Kepler. As far as he was concerned, what he was doing as an astronomer was akin to the work he had wanted to do as a theologian.

This faith also kept Kepler going through many difficult years to come. In 1598 the Catholic rulers of Graz suddenly ordered all Protestant teachers to leave the school, and though Kepler was allowed back for a short while, his position was untenable. In 1599 he traveled to Prague, where he had heard the Danish nobleman who was Europe's most famous astronomer, Tycho Brahe (1546–1601) had come in order to enjoy the patronage of the Holy Roman Emperor, Rudolf II (r. 1576–1612). As a nobleman, Brahe had been expected to take on a role in government, but instead he had petitioned the king to be allowed to be a practicing astronomer, after studying at the universities of Copenhagen, Leipzig, and Rostock and demonstrating a particular talent for astronomy. The king not only agreed, he also gave Brahe a land grant and provided financial support for him to build an observatory on the island of Hven, which became a research center, drawing natural philosophers from across Europe. In return, Brahe acted as an astrological advisor to the king. During the 1570s and 80s he made many important astronomical discoveries, including observations of a new star—a nova, which is an exploding star that temporarily becomes very bright—in 1572 where no star had ever been observed, challenging the Aristotelian and Ptolemaic principle that the heavens are perfect and unchanging. In 1577 his careful observations of a comet proved that it was moving through the orbits of the planets, which would be impossible if these orbits were embedded in the crystalline spheres that traditional Ptolemaic astronomy posited. He had also developed his own astronomical system, in which all the observable planets orbited the sun, which in turn orbited the Earth. This system has been largely forgotten, but at the time it was the most important competitor to the Copernican and Ptolemaic systems. However, his royal benefactor had died in 1588, and Brahe's relationship with the new king and his regency council (the new king was only eleven, so a group of councilors ruled in his stead for several years), steadily deteriorated, leading to Brahe's exile in 1597—which brought him to the court of Rudolf II in 1599, where he would meet Kepler.

When Kepler came to Prague to join Brahe, this was the beginning of an important relationship in Kepler's life, but it was a very rocky beginning to what would prove to be a very trying relationship. Brahe was a haughty nobleman and he treated Kepler like an impoverished nobody, which grated on Kepler. Nevertheless, though the two would work directly together for less than ten months total, and would disagree on many things, most importantly about the Copernican system, and fall into angry arguments more than once, Kepler and Brahe worked together off and on until Brahe's death in 1601. As with everything in his life, Kepler saw the hand of God at work in his relationship with Brahe, later writing: "God let me be bound with Tycho through an unalterable fate and did not let me be separated from him by the most oppressive hardships," Most importantly, Brahe was the best astronomer of his day and he had made many thousands of observations over the course of his lifetime. Upon his death he willed this data to Kepler so that he could complete a revised set of tables of celestial motion that came to be known as the Rudolfine Tables, after Tycho's patron at the end of his life, Rudolph II.

Kepler did complete the table, thought it took him many years to do so, but more importantly, he used the reams of astronomical data Brahe had compiled for his personal work. Kepler had never been able to make astronomical observations himself, due to his poor vision. If not for the mountains of data that Brahe willed him, he would have never been able to carry out the most important phases of his work. It certainly wasn't easy; throughout his life Kepler faced numerous challenges in his personal life. His wife fell ill with spotted fever and then all three of his children became ill with smallpox. One of his children then died, followed shortly by his wife. At one point, his mother was even arrested and tried for witchcraft; it was only Kepler's assistance that kept her from being executed. After 1618 the outbreak of the Thirty Years War in Germany—a conflict that began as a religious war between Protestants and Catholics before devolving into a struggle between the Hapsburg rulers of Spain and the Holy Roman Empire and the Bourbon rulers of France, which killed at least 1/3 of the population of Germany—kept Kepler and his family fearful and on the move during the last decade of his life, and at times he even fell under suspicion among Lutherans for presumed Calvinist sympathies.

Yet through all of these problems, Kepler always saw the hand of God at work, and as far as he was concerned, he found confirmation of this divine guidance everywhere in his astronomical calculation. In his *New Astronomy*, published in 1609 after a decade of hard work, he wrote of the problems of making mathematical sense of the data Brahe had left him: "Divine benevolence has bestowed the most diligent of observers, Tycho Brahe" on him, and thus "it is fitting that we accept with grateful minds this gift from God, and both acknowledge and build upon it." And build on it he did. Writing in the same work about his problems trying to calculate the orbit of Mars based on the observational data available:

> I was almost driven to madness in considering and calculating this matter. I could not find out why the planet would rather go on an elliptical orbit. Oh, ridiculous me! . . . With reasoning derived from physical principles agreeing with experience, there is no figure left for the orbit of the planet except a perfect ellipse.

In this way he was able to demonstrate mathematically that planetary orbits occur as ellipses, rather than perfect circles, but only after years of exhausting work. And why drive himself this way? As he explained in his highly influential *Epitome of Copernican Astronomy*, the most widely read textbook on astronomy of the seventeenth century:

> It is a right, yes a duty, to search in cautious manner for the numbers, sizes, and weights, the norms for everything [God] has created. For He himself has let man take part in the knowledge of these things . . . For these secrets are not of the kind whose research should be forbidden; rather they are set before our eyes like a mirror so that by examining them we observe to some extent the goodness and wisdom of the Creator.

In other words, Kepler never gave up on that notion he expressed to Maestlin in 1595, that his astronomical work was akin to theology. Driven by this belief he combined observational data with mathematical precision to build on Copernicus' work, replacing the perfect circles his Polish predecessor had insisted upon with ellipses, and proving to anyone who could work out the math that this *must* be the way planets actually move. He also made important contributions in other areas, such as the field of optics, and throughout his life he worked as an astrologer and even wrote a book about how to improve astrology. But it was his astronomical work that would prove to have the most long-lasting importance, and it was his religious faith that drove that work.

Moving away from Kepler's home in Germany, heading south into Italy, we find a land that is no longer quite as vibrant and wealthy in regards to European commerce as it had been during the Renaissance. Although cities such as Venice and Florence continued to trade across the Mediterranean in the seventeenth century, this trade no longer brought in the level of profits it once had. With the fall of Constantinople to the Ottoman Empire in 1453, Muslims held a monopoly on trade along the coasts of North Africa, Palestine, and Asia Minor, allowing them to charge high prices for the spices and silks that came from the east. That led European powers beginning with the Portuguese at the end of the fifteenth century to seek out and find a way around them, leading to an increasing flow of trade around the Horn of Africa instead of through the Mediterranean. Furthermore, the sixteenth century saw the Spanish conquering the civilizations of Central and South America, allowing them to plunder these peoples and bring the gold and silver of the new world across the Atlantic. Combined with the devastation wrought by wars between city-states, the invasion of the French at the end of the fifteenth century—which began the Italian Wars that lasted off and on from 1494–1559, devastating Italy economically—Italy simply wasn't the economic powerhouse it had once been. However, the city of Pisa would produce one more towering intellect, who would be one of the most important minds of the Scientific Revolution, in the sixteenth century: Galileo Galilee (1564–1642).

It was as a mathematician, astronomer, and physicist that Galileo would go on to make his mark on the world, and it was his approach to studying that natural world that sets him apart as the first true scientist in the Western world, even if that term and the cultural understanding of what it means to be a scientist wouldn't be developed until centuries after his death. But it was his problems with the Catholic authorities that have made him so famous that over the decades I've spent teaching, he's one of the few historical figures whom students enter into a freshman level class with an idea about who he was. And almost universally, those students—both those who are religious and those who aren't—believe that Galileo was a victim of the Catholic Church's anti-science stance. That's the view a great many historians have promoted, since the seventeenth century. Andrew Dickson White at least conceded that Galileo's prosecution "was not the fault of religion; it was the fault of the short-sighted views which narrow-minded, loud-voiced men are ever prone to mix in with the religion, and to insist are religion," yet he spends a dozen pages attacking efforts to limit the church's responsibility for Galileo's prosecution. For modern scientists, Galileo has become a hero, who stood up for what he believed against the repressive authority of the church, and, in the words of the chemist John W. Draper, led "the efforts of Science to burst from the thralldom [to the church] in which she was fettered." Carl Sagan in his book *Cosmos* casually states that Galileo "had been forced by the Catholic Church under threat of torture to recant his heretical view that the Earth moved about the Sun and not vice versa." Other scientists focus more on what Galileo represents for them, as Neil deGrasse Tyson does when he writes in his *Death by Black Hole* that the first time he looked through a telescope, he "communed with Galileo across time and space." In the same book, Tyson makes it clear that for him Galileo is a personal hero, one who "clearly distinguished the role of religion from the role of science," and who was "a rare exception among scientists," because he

> saw the unknown as a place to explore rather than as an eternal mystery controlled by the hand of God. As long as the celestial sphere was generally regarded as the domain of the divine, the fact that mere mortals could not explain its workings could safely be cited as proof of the higher wisdom and power of God.

I'm not sure who these people were who didn't attempt to understand the working of the heavens because these heavens were seen as a divine mystery not open to investigation; in reality, many medieval theologians wrote about the workings of the heavens. But the point is that for many, Galileo is not just a historical figure—he is a personal hero and a symbol of the repressive nature of the church.

That view of Galileo is so deeply ingrained and looms so large that there is a name for it among historians, "the Galileo legend." And like the best legends, it's constructed around a kernel of truth. However, like all legends, it doesn't describe reality very well and obscures more than it enlightens. So who was the historical Galileo and what are the details of his problems with the church? Born in 1564 to a father who was a musician and musical theorist of modest means, Galileo developed an early interest in not only his father's music, but also in the mathematics that underlay his musical theory. He also developed a lifelong need to chase funding, as his family lacked the wealth to provide him with a comfortable living and as the eldest of six children, his own family obligations would make his finances ever precarious. Perhaps just as importantly, his father was an iconoclastic innovator, not afraid to bend or even shatter rules in order to improve his craft, and he would pass this attitude on to his son. Furthermore, Galileo would learn to appreciate music theory—which is heavily mathematical in nature—and become an accomplished lutenist, thanks to his father's instruction. Given that numerous studies have shown that children who learn to play a musical instrument tend to develop a greater appreciation for and facility with mathematics than those who don't, his musical training was an important part of his development.

Galileo, as with many people of his day, was rather religious, so much so that his earliest ambition was to enter a religious order. This could be because his early education was at the Santa Maria monastery at Vallombrosa to the east of Florence, where he announced his intention to become a monk at the age of fourteen in 1578. His father had other plans and attempted to remove Galileo from the monastery in order to send him to the University of Pisa, but failed at getting a hoped-for scholarship, so his son stayed at the monastery until 1581. Although he would never become a monk (though he would become tonsured in 1616, when he was granted a pension from the papacy), Galileo's faith would inform his work for the rest of his life. In 1610 he wrote that "I give infinite thanks to God, who has been pleased to make me the first observer of marvelous things." He may not have viewed his work as a scientist in the same theological light as Kepler, but there is no doubt that his sense of religious wonder informed his interest in the study of the natural world.

Galileo did finally get into the University of Pisa in 1581, where his father intended that he study medicine. However, Galileo's interest in medicine was all but nonexistent. What really excited him was mathematics, and though his father attempted to keep him away from math, Galileo wouldn't be denied. The University of Pisa's curriculum was still heavily Aristotelian, and therefore mathematics was given very little attention there in the late sixteenth century. In fact, there wasn't even a chair of mathematics. However, Galileo convinced a friend of his father, Ostilio Ricci (1540–1603)—who was, at the time, the court mathematician to the Grand Duke of Tuscany—to give him private instruction. Ricci didn't view mathematics as a distinct field of study, but rather as a tool for solving practical mechanical and engineering problems, but he introduced Galileo to the work of Euclid (4th century BCE) and Archimedes (c. 287–c. 212 BCE), which would greatly influence Galileo. Both Hellenistic writers focused on geometry, and while Archimedes viewed the work he'd done in which he applied his mathematical skills to mechanics as beneath him, he showed how math could help one understand the physical world better. That was not the approach of most western natural philosophers throughout history, who viewed math as a metaphysical discipline that lacked any real connection to the physical world.

It's also likely that Galileo studied art while at the University of Pisa. More specifically, he not only read, but absorbed, *The Practice of Perspective,* published in 1596 by Lorenzo Sirigatti (d. 1625). This work included a description of the use of the art of chiaroscuro, a method for portraying three dimensions on a flat surface through a blending of light, shadow, and complex geometric forms. Renaissance artists developed perspective to a degree that it had never been used before in Western art, giving their works far greater depth and complexity than anything ever seen before. Galileo would become quite good at drawing in this style, but just as importantly, the study of art taught him how to see the world in new ways. When he looked at things near or far, it was with an eye trained to observe differences in shading and pattern in ways those untrained in the arts lack—which would prove to have enormous consequences in the future.

Galileo spent four years at the University of Pisa and two years studying math under Ricci's tutelage before he left the university without obtaining a degree. That's an important point to keep in mind, because Galileo never earned a university degree of any sort, and an M.A. was generally considered the minimum qualification necessary to teach at a university, which is still true today, in fact. He may have developed notions about the movements of a pendulum while at Pisa, though the story that he did so by watching the motions of a chandelier while timing its movement by using his own pulse as a timekeeping device is almost certainly apocryphal. Such a technique would hardly have been accurate enough to provide Galileo with meaningful data, and throughout his career one of the key factors in his approach to understanding the world is that he regularly developed new mechanical devices to provide ever-greater accuracy for his observations, as evidence by his first publication in 1586—a short tract called *The Little Balance* describing a device that could be used to weigh objects either in air or in water, based in large part on Archimedes' work. Regardless of what did or did not happen with the chandelier, it was at Pisa where Galileo became familiar with Platonic and Pythagorean philosophy, which would prove to have a tremendous impact on his career. Later, in 1619, Galileo would write:

> Philosophy is written in this grand book, the universe,
> which stands continually open to our gaze. But the book
> cannot be understood unless a person first learns to
> comprehend the language and read the letters in which it is

composed. It is written in the language of mathematics, and
its characters are triangles, circles, and other geometric
figures without which it is humanly impossible to understand a single word of it.

By the time Galileo left the University of Pisa he was a committed Neoplatonist and Pythagorean, who believed that mathematics—not logic, as the Aristotelians believed—was the language of
nature.

Between 1585 and 1589 Galileo worked as a private tutor and
public lecturer on mathematics in Florence and Siena. Galileo
was a tireless self-promoter and was fully aware of the truth of the
adage, "it's not what you know, but who." As such, he repeatedly
reached out to those whom he thought valuable to know, and this
technique worked. He made numerous connections, including
the German Jesuit Christoph Gau, known most commonly by the
Latinized version of his name, Christopher Clavius (1538–1612).
Clavius had worked on the development of the Gregorian calendar,
which replaced the older and less accurate Julian calendar across
most of Europe in 1582, and was the author of several influential textbooks on astronomy. Just as importantly, he established
contact with Guidobaldo del Monte (1545–1607), the Marquis of
Monte and the author of a commentary on Archimedes.

Like many of the best astronomers of the day, Clavius was
a Jesuit. Ignatius of Loyola (1491–1556) had founded the order
–officially the Society of Jesus—in 1540, with the explicit mis-
sion of providing education and using debate in order to bring
people to Catholicism. A soldier who came to religious life rather
late, after suffering an injury that ended his military career and
subsequently having what he considered to be a vision from
God, Loyola believed that the best way to combat heresy was not
by use of coercion, but through education, with the idea that
proper understanding would naturally lead to adherence to the
Catholic faith. Because education was central to the identity of
the Jesuits, many of the best scholars of the early modern period
were Jesuits, including astronomers. Clavius was one of these
astronomers, and he—like most astronomers of his day—adhered

to the Ptolemaic system even though he recognized that it had problems. This adherence had nothing to do with his religious convictions, but instead was a reasoned assessment of the available evidence. All observational data supported the idea that the planets (including the Sun) move around the Earth, and no one had been able adequately to explain why people standing on the surface of the Earth couldn't feel its motion, if it moved as Copernicus had claimed. Furthermore, the idea of an unmoving Earth that sits stably at the center of the cosmos was quite literally central to Aristotelian natural philosophy. Without a stable and unmoving Earth, Aristotelian physics no longer worked. We should keep these points in mind, for it's too easy to imagine that the heliocentric system is the only logical one, simply because modern people grow up being taught that it's the true and correct one. Clavius was a talented astronomer and mathematician who made advancements in math that are still recognized today, and his rejection of heliocentrism was a reasoned one, not faith-based.

In spite of such differences, Clavius was impressed with Galileo as a mathematician, and it was through his influence that Galileo received his first full-time teaching position as a mathematics professor at the University of Pisa in 1589. The money wasn't very good and Galileo was dissatisfied with his status as a lowly math professor, for in the sixteenth century such a position was far less prestigious than that of a scholar who studied ultimate truths, such as a philosopher or a theologian. Still, he needed the money and it was a start, but the low pay became even more of an issue when his father died in 1591 and Galileo was left in charge of his younger brother. This financial need is where Guidabaldo came in.

Galileo likely learned of Guidobaldo through the Marquis' writing on Archimedes. Guidobaldo's father had been a noted military commander and author of two books on military architecture, and it was through a combination of his activities as a soldier and a military architect that the Duke of Urbino had ennobled him, granting him both the title of marquis and the lands that went with it. His son had studied for a time at the University of Pisa, but with no need to earn a degree, had focused on mathematics.

He left the university when war broke out between the Ottoman Empire and the Holy Roman Empire in 1566 in order to fight the Turks in Hungary, and when hostilities died down with the death of the Ottoman sultan (from old age—he was seventy-two), Guidobaldo returned to his estates in order to write on the subjects of mathematics, astronomy, optics, and most importantly in the long run, mechanics (or what today we would call applied physics). Guidobaldo was part of the "Archimedean revival" of the late sixteenth century, using humanist techniques of close reading to develop Archimedes' physics in order to challenge the supremacy of Aristotelian physics. While teaching at Pisa, Galileo wrote a short tract on motion through a medium such as water, applying Archimedean principles instead of Aristotelian. In a very shrewd move and obviously in search of patronage, Galileo sent his work to Guidobaldo, who was extremely impressed. Over the years, Guidobaldo and Galileo would collaborate from time to time.

While at Pisa, Galileo was something of a problem. He felt himself to be cleverer than his traditional-minded—and better qualified—colleagues. And as far as the administration went, he was almost pathologically anti-authoritarian. He received fines on at least two occasions for missing classes, and perhaps as a bit of a payback he wrote a 300-line, extremely crude and intensely biting takedown of academia known as "Against the Donning of the Gown." Although he wrote the poem for his friends, it circulated throughout the university and would earn him a reputation as an iconoclast and a bit of an ass that would follow him for the rest of his career. Thus, he was only too happy when a better-paid position opened up at the University of Padua, and thanks to Guidobaldo's influence, Galileo became the new chair of mathematics. However, he received this position at the expense of Giovanni Antonio Magini (1555–1617). Magini had published numerous works, held an M.A. from the University of Bologna (where he'd beaten out Galileo for a job in 1588), and was, frankly, the stronger candidate—except for the fact that he didn't have the support of a wealthy and influential Marquis. Magini was justifiably incensed. Galileo was collecting enemies at least as rapidly as he was gathering supporters.

Over the next eighteen years Galileo taught at the University of Padua, but in spite of the increased pay he received, his money problems continued to worsen. Not only did he have to help support his younger brother and provide various other financial supports for his five siblings (including dowries for two of his sisters) that his role as head of the family incurred, but he also began living with a woman in Padua who would bear him two daughters and a son. Both daughters would go on to become nuns—one of his daughters, Virginia, who took the name Maria Celeste (1600–1634) upon her entry into the convent, likely helped prepare some of Galileo's manuscripts for publication— but Galileo would eventually recognize his son as legitimate and name him his heir. Much of what would transpire in Galileo's life was because of his need for money, which would drive him to pursue greater recognition and the increased funding he hoped would come along with that fame, since patronage from wealthy benefactors was still one of the primary ways for intellectuals to garner an income during this period.

While at Padua, Galileo made and sold mathematical instruments, but also began to work more seriously on problems of motion, using Archimedean principles, and wrote of experiments using inclined planes to test the rate of falling objects in order to demonstrate that objects of varying weights do not fall at different speeds, all other things being equal—which directly contradicted Aristotelian physics. It was also at Padua when he first expressed an interest in Copernican astronomy, when he wrote to a former colleague at Pisa to defend against a mistaken critique of the system in 1597. Word got around, and Johannes Kepler wrote to Galileo before the end of the year, sending him a copy of his first published work and urging him to support Copernicanism openly. However, this is where things get a bit complicated, for Galileo had become interested in Copernicus' theories for the wrong reason. Rather than being drawn to it because of any inherent benefits it might provide, or even due to philosophical considerations as Copernicus and Kepler had been, Galileo felt that the idea of a moving earth supported a theory that was important to him, one about the tides.

Galileo was very interested in gaining the attention of wealthy potential benefactors, and there were two related ways available to him to do so—by stirring up public controversies and by demonstrating how the work he was doing had direct applicability. Since most of Italy's wealth was drawn from seaborne commerce, Galileo sought to demonstrate how his evolving understanding of physics could help better people's understanding of the sea. From reading Copernicus, Galileo had come to believe that a moving Earth could explain the tides. To provide a very simplistic explanation of how he thought this might work, simply imagine a bucket with water in it, which gets swished one way and then the other. The water within the bucket slops up (and probably over, depending upon whether your imaginary bucket is as full as mine) one side and then the other as the bucket swings around. Galileo believed that the motion of the Earth affected the oceans in a similar fashion. This idea would consume Galileo for much of his life, leading him to write a treatise on it in 1616 and he wanted to make it the central theme of his most important work in 1632 (more on that later), but when he wrote of this idea to Kepler, the response wasn't positive. Kepler objected to the idea that the motions of the Earth caused the tides for the very good reason that scholars had known for centuries that the Moon causes tides, even if no one really knew exactly how. Galileo, however, didn't like being contradicted so this exchange spelled an early end to any scientific collaboration that otherwise might have developed.

During his years at the University of Padua between 1592 and 1610 Galileo did vitally important work that would require a lengthy treatment, were this a history of science. Although his income had tripled when he moved on from Pisa, his need to provide a dowry for is sister in 1601 and the costs associated with first a live-in mistress, then the three children she would give birth to, meant that Galileo was pressed for funds, and the only way he could solve that problem was by gaining patronage—which required that he increase his name recognition, and expand his status beyond that of a mere mathematician and into the realm of philosophy. Therefore, he sought to investigate the causes of things while challenging Aristotelian methods of explaining the world. This included both thought experiments as well as actual experiments, such as gathering data by rolling balls of varying weights down inclined planes, timed against a water clock, in order to test his hypothesis that Aristotle was wrong, that bodies of different weights actually fall at the same rate, if other factors are equal. However, for our purposes the most important elements of his work was his continued efforts to draw closer to wealthy and powerful patrons, and his last years at Pisa when he shifted his focus toward the heavens.

Galileo's big break in his quest for patronage occurred in 1605, when the Grand Duke of Tuscany invited Galileo to use his vacations from teaching to tutor his son—who would one day become the Grand Duke. The Grand Duke, Ferdinand I (r. 1587–1609) was intensely interested in natural philosophy and he wanted to ensure his son was properly educated. Galileo had been pushing to become a tutor to the boy since 1601, so he leapt at the chance. This position would be of immense importance, because it meant that Galileo was in a position to profit handsomely when his teenaged pupil became the new Grand Duke, Cosimo II (r. 1609–1621) in 1609, at the ripe old age of nineteen, when his father died somewhat unexpectedly at fifty-nine. This was not an age when people were guaranteed a long life—Cosimo would die at thirty from tuberculosis.

Upon Cosimo's accession to the dukedom in 1609, Galileo petitioned to become his court mathematician, but he simply lacked the prestige necessary to attain such a lofty position. However, his financial demands were becoming increasingly pressing, so he needed this position. He needed to do something truly bold in order to increase his prestige, and in that same year he learned of a means by which he might do so—the telescope. A pair of Dutch inventors had developed the device in the previous year, and had attempted to sell one to the Venetian Senate. They declined, so Galileo learned how the device was built and more importantly, how to improve upon it. He gifted his new and improved version of the telescope to the Venetian Senate, which controlled Padua. Pleased by the gift, the Senate increased his salary.

But more importantly, Galileo turned his own telescope to the heavens, and what he saw there was heavily informed by his own training in art. When he looked at the Moon, instead of seeing the smooth, perfect surface that Aristotelian and Ptolemaic ideas about non-terrestrial substances said should be there, he saw hills, valleys, plains, and giant holes in the ground. While anyone who has looked at a full Moon with the naked eye might be surprised by the idea that people once thought of the Moon's surface as perfect and unblemished, we should remind ourselves that people see what they expect to see, and according to Aristotle and Ptolemy the Moon—like all heavenly bodies—is made of a fifth element, quintessence, and is therefore perfect. Any apparent blemishes are artifacts of the atmosphere, the distance, or defects of our vision. But trained as an artist to understand perspective, Galileo recognized right away that what he was seeing was a surface that looks a lot like that of the Earth's in many ways, meaning that the Moon is not made up of quintessence. It's made of the same stuff the Earth is, and therefore the same laws of physics that work on the Earth should apply to the heavens. Turning his telescope away from the Moon, Galileo saw thousands of stars where none appeared to exist to the naked eye, suggesting that the universe is much larger than had been expected. Finally, and most importantly, he saw four new planets—moons actually, orbiting Jupiter—which he dubbed the "Medicean Stars," after the Medici rulers of Florence, in a bid for patronage.

Galileo published his findings the following year in a book titled *The Starry Messenger*, and overall it was more a sensationalist work than one that was paradigm shifting. The moons of Jupiter did raise questions about the Ptolemaic system in which everything in the cosmos orbits the Earth, but there was no reason that such findings couldn't be accommodated to traditional astronomy. And it fit perfectly well with the Tychonic system, which posited that everything except the Earth orbits the Sun, which in turn orbits the Earth. Galileo does mention within the work that he will produce a follow up that will prove that the Earth moves, but that's barely a throw-away line that garnered little attention. However, the book as a whole generated a great deal of conversation—and ended in Galileo being called to Florence to be the new Court Mathematician and Philosopher to the Grand Duke of Tuscany.

One important point to note before we consider the next phase of Galileo's work is in regards to the religious reception of *The Starry Messenger*. There was some initial pushback against Galileo's view of the moon (accompanied by his personal drawings) as an imperfect body similar to the Earth, but that pushback was generated primarily because not all observers understood what they were seeing through a telescope the same way Galileo did—many lacked his artistic training and saw what the Aristotelian view of the heavens predisposed them to see. Many have seen the controversy as a religious one, beginning with Draper who wrote: "These and many other beautiful tele-scopic discoveries tended to the establishment of the truth of the Copernican theory and gave unbounded alarm to the Church. By the low and ignorant ecclesiastics they were denounced as deceptions or frauds." It's not at all clear where he would have gotten such a notion, however. By the end of 1610 Christopher Clavius—the Jesuit astronomer— had confirmed Galileo's find-ings—including the discovery of the moons of Jupiter— with a telescope in Rome, which he reported directly to the pope who was impressed, rather than angered.

At first things went very well for Galileo in Florence. Continuing to use his telescope, he discovered that Venus has phases just as the Moon does. That was much more important, in many ways, than what he'd previously published in *The Starry Messenger*. In the Ptolemaic system, Venus simply couldn't have phases—the only way it could have phases is if it is sometimes between the Earth and the Sun and sometimes on the opposite side of the Sun from the Earth. This bit of data seems to have ended any doubts Galileo might have had about the validity of Copernicanism, but there was a problem—the data also fit perfectly with the Tychonic system. In fact, the two systems are observationally equivalent— *no* observation Galileo could make would fit with the Copernican but not the Tychonic system. Really, the only thing to push Galileo more toward Copernicus' model was his Neoplatonic convictions and his certitude that the Earth moves, but he had not observational data to back that up. In fact, the observational data contradicted the idea of a moving Earth, for as Clavius pointed out and Galileo well knew, if the Earth orbits the Sun, that should mean that the positions of the fixed stars appear to shift as the Earth moves, a phenomenon known as the stellar parallax—but try as he may, Galileo couldn't demonstrate the existence of such a parallax. Telescopes of the seventeenth century simply weren't good enough, and it would be 1838 before a German astronomer finally observed the stellar parallax. However, there was no reason for anyone in the seventeenth century to fault the powers of the telescope—it appeared the one prediction that would fit with a moving Earth had failed. However, in spite of this failure, Galileo was riding the crest of a string of successes, and by the end of 1611 Galileo was invited to visit Pope Paul V (r. 1605–1621) in Rome, where he was received as a guest of honor by the pope and a number of cardinals. If I can be allowed a bad pun, his star was in the ascendant.

However, as that was happening, there were hints that trouble loomed. In 1611 a philosopher named Ludovico delle Colombe (1565–after 1623) wrote a work titled "Against the Motion of the Earth," that used Aristotelian arguments to attack Galileo, but ended with an aside that the notion of a moving Earth contradicts the bible. And in the following year, in 1612, a Dominican priest who couldn't even get Copernicus' name right (he called him "Ipernicus") mentioned to Galileo that heliocentrism was contrary to scripture. Concerned that conservative forces might rally against him, Galileo wrote a short letter in 1613, which he expanded in 1615 as the "Letter to Madame Christina of Lorraine, Grand Duchess of Tuscany, Concerning the Use of Biblical Quotations in Matters of Science." In this letter, Galileo wrote reasonably enough that "I do not feel obliged to believe that the same God who has endowed us with sense, reason, and intellect has intended us to forgo their use," which is perfectly consistent with traditional Catholic theology. However, he went on to lay out a full-throated statement of his acceptance of heliocentricism as a physical reality:

> I hold the sun to be situated motionless in the center of
> the revolution of the celestial orbs while the earth rotates
> on its axis and revolves about the sun . . . these discover-
> ies clearly confute the Ptolemaic system, and they agree
> admirably with this other position [of Copernicus] and
> confirm it.

That *might* not have been such a problem, but he then went on to discuss passages from Joshua 10:15, where at Joshua's command (but through divine power), "The sun stopped in the middle of the sky and delayed going down about a full day," giving Joshua time to win against the Amorites. Galileo appealed to the fifth century theologian Augustine in order to argue that this passage should be read figuratively, rather than literally, and provided a scientific discussion about the motion of the Sun in order to show why it must be taken this way.

Galileo's position was perfectly consistent with medieval views of biblical interpretation. However, in the wake of the Reformation the Council of Trent had decided that the bible should be interpreted literally wherever possible. But more importantly, Galileo was a mathematician—and one with no university degree, it should be remembered—who was telling theologians how they should interpret the bible. Galileo had friends who were theologians who warned him against this tactic, and he really should have listened. There's no better way to anger an expert than for a novice to tell him or her how their job should be done. Although written as a personal letter to the Duchess Christina, Galileo intended for this work to be public knowledge and in fact sent a copy of it directly to the papal inquisition at Rome. The result was to intensify the anger of those who already didn't like Galileo, so he decided to travel to Rome and meet with the inquisitors himself. He had already been in correspondence with the pope's chief advisor on matters of theology, Cardinal Robert Bellarmine (1542–1621), so in 1616 Galileo traveled to Rome to meet with him. The two had a very productive meeting, for the most part, and agreed on most things—such as that true knowledge and the bible could never conflict—but disagreed over the status of hypotheses. Bellarmine saw such positions as assumptions used for the sake of convenience, which didn't necessarily correspond to physical reality (which is how theologians used the term hypothesis), whereas Galileo saw them as assumptions likely to one day be proven.

The key, though, is that Bellarmine pointed out to Galileo that he had no evidence for the motion of the Earth, which, frankly, was completely accurate. In fact, the one bit of evidence that should have existed *if* he were correct—the stellar parallax—seemed not to exist. Galileo disagreed, believing that his theory of the tides proved the motion of the Earth, but Bellarmine (like every other educated European of his day) knew that Galileo was wrong about the tides. Instead of pressing that point, he simply told Galileo that his interesting hypothesis about the tides was unproven and therefore couldn't act as evidence to support the motion of the Earth. Therefore, he could not hold or defend Copernicus' helio-centric view *as proven*. While from a modern viewpoint, we might react negatively to the idea of a religious leader telling a scien-tist (and I have argued elsewhere that Galileo really was the first European to do what we now refer to as science), we should keep in mind that in the seventeenth century most universities were run by the Church, which was far more involved in education and printing than most people today would deem appropriate—but which was perfectly normal for the period. So while I, as a mod-ern academic, would be enraged to find myself or a colleague in the same situation Galileo was in with Bellarmine, to judge this event by such an ahistorical standard is to miss the point and misunderstand what actually happened. Thus, what Bellarmine was doing was telling Galileo that he couldn't willy nilly make pronouncements about the way the world works without having suitable evidence to back such pronouncements up, which he clearly lacked.

This is a complicated issue, and there were certainly elements of conflict between religious belief and the problems Galileo found himself. In 1616, the Sacred Congregation of the Index also con-demned Copernicus' *On the Revolutions* for the first time, at least until such time as it could be corrected. However, the point is that Galileo's problems with the church were far more complicated than a simple clash between faith and reason, religion and science. Conflicting personalities, Galileo's ardent sensationalism, and the simple fact that Galileo was so often wrong while continuing to aggressively attack those who disagreed with him—all of these issues and more played a part in his story.

Regardless of how any of that sounds to a modern person, Galileo understood political realities, so between 1616 and 1623 he publicly stated that the Copernican system was only a "hypothesis" in the sense that Bellarmine used the term, an occasionally useful fiction, while at the same time rejecting both the Ptolemaic and Tychonic system out of hand whenever the opportunity presented itself. In short, his public position was that no one had presented an adequate description of the structure of the cosmos. However, this didn't mean that he stayed out of trouble. Instead, between 1619 and 1623 he engaged in a rather pointless yet nonetheless heated dispute with Orazio Grassi (1583 -1654) a Jesuit astronomer, and his colleagues over the nature of comets. Grassi, a student of Christopher Clavius, writing under the pseudonym Lathario Sarsi, worked with colleagues to write an *Astronomical and Philosophical Book* about Tycho and Kepler's proofs that comets are wandering "planet-like bodies" that move through what we would now call the Solar System. Galileo reacted in the strongest possible terms, asserting that comets are atmospheric phenomenon and calling into question the intellectual and scientific competence of anyone who disagreed into question.

The key point here is that Galileo was wrong about comets and Grassi—and his fellow Jesuit astronomers—were not only right, but they also had the data and mathematics to prove it. By engaging in this argument, Galileo made enemies of most of the best astronomers of the day while demonstrating what seemed to almost everyone an amazingly poor grasp of astronomy. We must keep these facts in mind if we are to understand the Galileo affair.

In 1623 Pope Paul V died and the College of Cardinals elected Maffeo Barberini to be Pope Urban VIII (1623–1644). Urban was a highly intelligent and sophisticated man who had earned his doctorate in law from the University of Pisa in 1589. He had previously interacted with and sometimes supported Galileo in his many academic disputes, and the two viewed one another as friends. Very soon after Urban's election, Galileo traveled to Rome in the Spring of 1624 and had a series of conversations with the pope. He asked for permission to write a book on heliocentrism, and Urban agreed that he could do so. However, he insisted on two things: Galileo could not present Copernicanism as proven and he couldn't focus his work on the tides, which was what Galileo wanted to do. Again, Urban pointed out that Galileo's theory of the tides simply lacked support.

It took Galileo a few years to complete his book. His eyesight was failing, and he encountered personal problems such as a wastrel nephew whom he was supporting in Rome who stirred up controversy by his blatant irreligiosity. However, in the end Galileo published his masterpiece, *Dialogue on the Two Chief Systems of the World*, in 1632. He wrote this work in Italian to expand its reach and in dialogue form to make it easier for more people to understand the points under consideration. The dialogue was between three figures, two of whom were named for close friends who defended Copernicanism, and a third who proved the most problematic, who defended Ptolemy and traditional Aristotelianism. He named this last figure after a Roman writer who had written on Aristotle, Simplicio. By using the mechanism of a discussion, Galileo seems to have truly felt that he was avoiding any conflict with the order that he not present the Copernican system as proven, but there were a number of problems with his presentation.

First of all, no person reading the text can walk away from it without feeling that Simiplicio lost the debate, even though he closes with words taken almost verbatim from the pope's mouth regarding divine power. To make matters worse, throughout most of the book Simplicio comes across as an idiot—which is, in fact, what "simplicio" suggested to seventeenth-century speakers of Italian. So by placing words the pope had spoken into Simplicio's mouth, it appeared to more or less everyone that Galileo was calling the pope an idiot, and it certainly seemed as if he was ignoring the pope's orders not to present the Copernican system as proven. Galileo had spent decades making enemies among the best astronomers of the day, who almost to a man were Jesuits (just as the pope was), so they were only too eager to bring these matters to the pope's attention.

To make matters almost infinitely worse, Pope Urban VIII was himself in a great deal of trouble with his own College of Cardinals. He had come to office during a very difficult period for the papacy. The Thirty Years War was raging across Europe, and some within Italy saw this as a good time to encroach upon papal territory. Urban, however, wasn't going to allow that to happen on his watch, so he borrowed and spent vast sums fighting wars in Italy to increase the size of the Papal States, supporting one side or another in the Thirty Years War, and acting as a patron for artists and intellectuals in order to boost the prestige of the papacy. The result was that by the 1630s he had drastically increased the papacy's debts and the Spanish members of the College of Cardinals were becoming downright mutinous. He couldn't afford to look weak and needed to do something bold to show his strength. Therefore, driven by a complicated set of emotions, ranging from a personal feeling of betrayal from a man he had called friend to fear for the success of his papacy, Urban commanded the papal inquisition to investigate the case. Publication of the *Dialogues* was suspended and Galileo was ordered to appear in Rome before the inquisition.

What followed is a sad affair that has been made worse in the popular imagination by poor or ideologically driven history. Galileo was seventy and had failing eyesight and asked to be excused from traveling to Rome, but was told that he would be convicted *in absentia* if he didn't attend, so off he went. Historians and those interested in history from the eighteenth century to the twentieth have written that "Galileo . . . groaned away his days in the dungeon of the Inquisition," as Voltaire put it in 1728, and was thrown "into a Catholic dungeon and threatened with torture," as Carl Sagan put it in 1981. This view of Galileo was no doubt what that neuroscientist Sam Harris had in mind when he wrote in 2005 that the Catholic Church has had a habit of "torturing scholars to the point of madness for merely speculating about the nature of the stars" during past centuries, as he states in his book *The End of Faith*. The reality is far less dramatic. Rather than being locked in a dungeon, thanks in no small part to the influence of the Grand Duke of Tuscany who treated this entire affair as a matter of state, while in Rome Galileo stayed at first at the Tuscan embassy, then at the prosecutor's six-room apartment, and was allowed a servant who prepared him two meals a day. The inquisitors did threaten Galileo with torture at one point, but at no point was this threat carried out.

In the end, Galileo took a plea bargain, and admitted to having written his book in such a way as to give readers the erroneous impression that the Copernican system was proven, but denied that this had been his intent. In turn, the inquisitorial tribunal agreed to press only the lesser charge of failing to heed the warning not to teach the truth of Copernicanism. Galileo was pronounced guilty of "vehement suspicion of heresy," an intermediate crime, and the sentencing document noted that the beliefs he had taught that went against the teachings of the Church were that the Earth moves, and that the bible is not a scientific authority. The church banned the *Dialogue* and released trial documents stating that information had been obtained from Galileo by "rigorous examination," which often meant torture, and closed that he had been condemned "to formal imprisonment." Urban VIII released this version of events on purpose, to make him appear to be a decisive and forceful opponent of heresy.

Thus, it makes sense that for centuries most people thought Galileo had been imprisoned and tortured. However, cracks began to appear in this account when in 1774/5 researchers uncovered documents demonstrating Galileo spent the last nine years at his villa in Arcetri, where his daughter Maria Celeste came to take care of him, until she grew ill and died. In Arcetri, rather than being kept in rigorous confinement, he was visited by a steady stream of students and famous men who wanted to meet him, including the English philosopher Thomas Hobbes (1588–1679) and the English poet John Milton (1608–1674). Still, it would be 1867 before the minutes of the inquisition's interrogation of

Galileo emerged, which showed that Galileo was never tortured. During the last years of his life, Galileo put together one more book, though he had been forbidden to publish further, which he gave to a former pupil who had it published in Holland. He puttered around with experiments, met with former students and friends, and played the lute until he died in 1642.

I've spent more time dealing with Galileo than any other figure in this book, because for many he is *the* symbol of the inherent conflict between religion and science. And there is a kernel of truth to that, as the Catholic Church struggled with the forces the Reformation and the Wars of Religion unleashed on Europe. However, to present Galileo as the victim of repressive religious beliefs and an oppressive Catholic hierarchy is massively simplistic and misses almost all of what makes the historical person of Galileo interesting. He was combative, insulting, and lacked the academic credentials of his opponents. And he was glaringly and obviously wrong about significant points of fact, such as the nature of tides and comets, in ways that were obvious to his opponents, who could at least in regards to comets show overwhelming evidence of Galileo's mistaken views—only to be met by ridicule and insult. There's no doubt that Galileo was a genius (he certainly never doubted it) but the resolution of the "Galileo Affair" had at least as much to do with his character flaws and need for fame as with questions regarding religion. Furthermore, the harm that came from this affair—in its aftermath fears that one might fall afoul of the Catholic Church while investigating nature seemed justified to many—can be blamed on Galileo at least as easily as blame can fall on Urban VIII or Cardinal Bellarmine. If he had written a more measured book, as he had promised to write, there would never have been a "Galileo Affair."

In the end, it's clear from even this cursory examination of figures such as Johannes, Kepler, and Galileo, that it's far too simplistic to accuse religion of standing in the way of science in the sixteenth and seventh centuries. After all, both were deeply devout men—though it's probable that Kepler's faith was more intense than Galileo's—both were driven to study the natural world at least in part by their desire to understand the beauty of what they saw as God's creation, and for Kepler that was the central motivation. However, the facts of their lives show that the terms of the problem were coming to be more complex. Kepler was a Lutheran and Galileo was a Catholic. In the centuries to come, that complexity was only going to increase as the nature of what it was to be a Christian fractured further, and by the 18th century we have the emergence of those who deny that a deity is actively involved with human life—and some who deny the existence of a creator deity at all.

Further Reading

Christianson, John Robert. *On Tycho's Island: Tycho Brahe, Science, and Culture in the Sixteenth Century.* Cambridge: Cambridge University Press, 2003.

Drake, Stillman. *Galileo at Work: His Scientific Biography.* Mineola: Dover Publications, 2003, 3rd printing.

Finocchiaro, Maurice A. "Myth 8: That Galileo was Imprisoned and Tortured for Advocating Copernicanism. *Galileo Goes to Jail and Other Myths about Science and Religion.* Ron Numbers, ed. Cambridge: Harvard University Press, 2009, 68–78.

Lindberg, David C. and Robert S. Westman. *Reappraisals of the Scientific Revolution.* Cambridge: Cambridge University Press, 1990.

Lipking, Lawrence. *What Galileo Saw: Imagining the Scientific Revolution.* Ithaca: Cornell University Press, 2014.

Mayer, Thomas F. *The Trial of Galileo, 1612–1633.* Toronto: University of Toronto Press, 2012.

Porter, Roy and Mikuláš Teich. *The Scientific Revolution in National Context.* Cambridge: Cambridge University Press, 1992.

Renn, Jürgen. *Galileo in Context.* Cambridge: Cambridge University Press, 2001.

Scotti, Dom Paschal. *Galileo Revisited: The Galileo Affair in Context.* San Francisco: Ignatius Press, 2017.

Sobel, Dava. *Galileo's Daughter: A Historical Memoir of Science, Faith, and Love.* New York: Bloomsbury, 2011, 2nd edition.

Thoren, Victor E. *The Lord of Uraniborg: A Biography of Tycho Brahe.* Cambridge: Cambridge University Press, 1990.

Voelkel, James R. *Johannes Kepler and the New Astronomy.* Oxford: Oxford University Press, 1999.

Westmann, Robert S. *The Copernican Question: Prognostication, Skepticism, and Celestial Order.* Berkley: University of California Press, 2011.

CHAPTER 5

The English and French Scientific Revolutions

T he Italian decline in economic and political importance didn't mean that Italians were no longer important to the early modern European economy. Thanks to the long importance of Italian seafaring merchants, many of the geographic discoveries—at least they were discoveries as far as the Europeans were concerned—were made by Italian captains, such as Christopher Columbus (1451–1506) and Amerigo Vespucci (1451–1512, after whom "America" is named). Furthermore, Venice would continue to play an important, though declining, role in trade through the seventeenth and eighteenth century. However, the balance of power was definitely shifting. As the only European nation not actively at war in the early fifteenth century, with a ruling family that recognized the small size and poverty of their nation, the Portuguese had managed to find a route to China and the Spice Islands that bypassed the Ottoman Empire by going south around the Horn of Africa by the end of the fifteenth century. Spurred on by this development, and by Columbus' voyages, English captains had sailed eastwards in search of a new route allowing access to the riches of the East, and instead had found North America. By the seventeenth century the American colonies were providing wealth for the English in the form of tobacco and the pelts of animals such as beavers, which were known as "soft gold" due to their value. The French likewise accessed the pelts of North America that were increasingly valuable during the Little Ice Age (1645–1715). Meanwhile, the Spanish had conquered the Aztecs and the Incas, gaining access to actual gold and silver; they squandered this wealth on one war after another, but the introduction of these precious metals into the European economy allowed for economic expansion for the economic elites, even as the explosive inflation it caused did great economic harm to Europe's laborers. The primary beneficiaries of all these changes, at least from the perspective of the development of science, were the English and to a lesser extent the French.

We'll begin our story with the English. During the sixteenth and seventeenth century England had gone through a wide range of changes that would forever alter the nation. Henry VIII (r. 1509–1547) had sought to increase his power over England. That, plus a desire for the male heir that his first wife could not provide, led him into a conflict with the Pope who refused to grant his divorce from a woman who was the aunt of the Holy Roman emperor—whose armies sacked Rome in 1527, about the time when Henry asked for this divorce—that culminated in England's break from the Catholic Church, and the creation of a new state church in England, the Church of England. During this same period, the largest and wealthiest landowners in England were gaining ground against their more modest neighbors, as inflation bit into the profits of farmers and it became increasingly clear that raising sheep for their wool held greater economic promise than farming. Thus, large landholders bought up farms from their less wealthy neighbors, and enclosed much of this land with fences in order to raise sheep for their wool. Since such pastoral activities required fewer workers than farming, many farmers and laborers were cast off their lands, contributing to the ranks of the growing number of impoverished vagabonds who roamed England, putting a fright into the hearts of the economic elites.

In this atmosphere of religious, economic, and social uncertainty, new religious interpretations emerged in spite of royal efforts to stop this from happening. The largest group of such people were influenced by the teaching of Jean Calvin (1509–1564), the

French lawyer turned religious reformer whose theology centered on the notion that all people were born predestined for either heaven or hell, and nothing a person did could alter his or her fate. Thus, it was important to read the bible and place one's faith in God's divine goodness, in hopes of being among the handful of the elect instead of those damned to hell. As Puritanism developed in England, many of these Calvinists sought signs of whether or not they were among the elect. Ministers argued that such signs could be found in outward success, giving Puritans a drive to succeed economically that would lead many of them to find success in business in ways that were beneficial to the English economy. Although landed wealth was where the highest level of social status was to be found, the landed aristocrats weren't blind to the profits being made by merchant traders, and weren't shy about investing the profits they earned from the wool markets in the burgeoning stock market (a seventeenth-century innovation) or directly with the emerging middle class of merchants and bankers in England. Thus, the English landed and mercantile classes enjoyed the benefits of a rapidly growing economy.

However, all did not go smoothly in England during these years. Many resisted Henry VIII's religious reforms, and many more were dissatisfied with their declining economic status as enclosure drove large numbers of families from the land, causing rebellions such as the Pilgrimage of Grace (1536–37). Henry quickly crushed such resistance at home, in order to focus his attentions on expensive and ultimately unsuccessful military adventures abroad. This may have been a boon for the English in the long term, though, as it meant that he squandered the wealth his father had spent years amassing as well as the wealth he took (some might say "stole") from the Catholic Church, when he closed monasteries across England and took church lands as his own—at least until he sold these lands to his nobles for quick cash. Therefore, he didn't use the wealth at his disposal to do any of the things the French monarchy was doing, such as building up a standing military or otherwise enhancing the instruments of power available to the monarchy. Meanwhile, he leaned heavily on Parliament to accomplish his various divorces, the split with the Catholic church, and other elements of his domestic policy—which ultimately left members of Parliament with a growing confidence and a sense that Parliament deserved to be a partner in running the country. Thus, it was only the skillful governance of heirs such as Queen Elizabeth I (r. 1558–1603) and James I (r. 1603–1625), who deserves far more credit as a statesman than he is usually given, that kept England from lapsing into conflict or becoming the victims of foreign aggression.

Not all monarchs had equal skill and talents, though. Upon the death of James I, his son came to the throne as King Charles I (r. 1625–1649). Although he was intelligent and hardworking, Charles also suffered from a dangerous lack of self-confidence (no doubt brought on by his father, who always called him "baby Charles" and often made fun of Charles' stutter—early modern parenting wasn't always big on building up a child's self-esteem), which he overcompensated for by lashing out at anyone who disagreed with or questioned him. He was also autocratic by nature, buying into the idea of divine right monarchy that was so influential on the Continent, especially in France. His wife, Henrietta Maria (1609–1669), was a French princess, who had grown up in a court where the power of the king was more or less absolute, so she only encouraged such beliefs in her husband. Charles became king during a time of both rising religious tension in England, as Puritans increasingly demanded greater legal rights, as well as growing economic tension, all while England faced a range of international troubles. Thus, as a result of Charles' ineptitude as much as anything else, England fell into a civil war that pitted Parliament against king, which was greatly complicated by its relationship with Scotland and Ireland. That's why many refer to this war as the War of the Three Kingdoms (1642–1646), which killed off almost 12% of the population of England, Scotland, and Ireland. When we consider that the enormously bloody American Civil War (1861–1865) killed "only" 2–3% of the population of the United States, it's clear that this was a desperate conflict indeed. The war ended with Charles a prisoner, but too many issues left undecided, so with his escape and flight to Scotland (he was the son of James I, who had been King James VI of Scotland before taking the English throne, after all), there was a brief convulsion of renewed warfare in 1648, which ended with Charles' decisive defeat. He was then tried for treason and executed in 1649, the first and last time a sitting monarch in England would ever be publicly executed.

The English Civil Wars were bloody, devastating, and destructive. However, there were, as always, winners and losers. Land speculators earned fortunes by buying up grants of Royalist lands made to Parliamentary soldiers in place of the pay Parliament could never quite come up with for them. Furthermore, Parliament was left in a permanently strengthened position. Defeating and killing a king can do that for you, it seems. And even as war raged, many wealthy families went on with the serious business of increasing their wealth, and with peace, that would come to be the universal past-time of the English.

One such family was that of the Boyles. Richard Boyle (1566–1643) was a fairly common sort of figure in England of the day; as the second son of a well-established family, he would inherit very little beyond a good name, a good education, and a small purse full of coins. However, he would parlay this inheritance into a virtual empire comprised of massive tracts of land in Ireland, which was a colonial holding of the English crown at the time. He then leveraged these holdings to gain a peerage, being created the first Earl of Cork and Viscount Dungarvan in 1620, as well as gaining extensive English lands in the counties of Dorset and Somerset. But of all the successes Richard had, none could equal that of his seventh son and fourteenth child (Richard was a very, very busy fellow), Robert Boyle (1627–1691), born in Lismore Castle on his father's estates in Ireland.

As a boy, his father saw to it that Robert received the best education possible, due in no small part to the fact that the family's pedigree was somewhat newly minted. Thus, in order to gain the respect Richard felt the Boyle family name deserved, he would have to make sure he owned the right sort of residences, filled with the right types of artwork, and peopled with Boyles who had the right sort of education. Thus, from an early age Robert had tutors who taught him languages such as Latin, Greek, French, and even Irish, which was a very unusual language to learn even for the landed Anglo-Irish gentry at this date. His facility with languages meant that he was able to read Diogenes Laertius' (180–240) *Lives of the Philosophers* in the original Greek, giving him an early interest in Epicurus' ideas about atomism. When his mother died, Robert was sent to Eton College at the age of eight. King Henry VI (r. 1422–1461, and 1470–71) had founded this boarding school in 1440, and ever since, it has been the training ground for many of the movers, shakers, and rulers of England. At Eton it's likely that Robert's education would have been a mixture of classical and humanist subjects, with an emphasis on Latin and Greek, but also a grounding in math and mathematical disciplines such as physics and astronomy. At twelve, his father sent Robert and his next oldest brother, Francis (1623–1699), the future 1st Viscount Shannon, to continental Europe with a tutor.

One of the places Robert would spend time in Europe was Geneva, Switzerland, where a tutor taught him and his brother "the polite arts," that is, how to behave like gentlemen, as well as subjects like mathematics. It was in Geneva where the thirteen-year-old Robert Boyle would undergo a serious, and lasting, conversion experience. One night thunder awoke him during a powerful storm that raged outside. Trembling and, according to his later accounts, fearful for his life—and we should remember that before Benjamin Franklin (1706–1790) invented the lightning rod, it was not uncommon for lightning to destroy buildings—the young Robert Boyle did something that many people have often done when frightened; he prayed and begged God to protect him, promising to devote his life to serving Him if he survived the storm. However, the difference between Robert Boyle and so many others who make such bargains is that he meant what he said. From that point on, he was not only a committed but also a fervent Christian, and it is only within that context that his life's work can be understood.

Although his father was a staunch Royalist, Robert Boyle would be much more Parliamentarian in his sympathies, though he would never take up arms for either side. However, it's likely that his Parliamentarian sympathies would shield the family from the backlash against Royalists that would later occur. Returning in 1644, his older sister Katherine introduced him to the physician and general polymath Samuel Hartlib (1600–1662), who had immigrated to England from Germany in 1628. It was Hartlib who inspired Boyle to have greater interest in alchemy than astronomy, and in the 1640s Boyle attempted to get a furnace and other tools necessary to alchemical work. However, none were available, so he spent the next few years doing intense reading, of both classical and current authors of natural philosophy. This time was far from wasted—Boyle never attended a formal school, but thanks to the tutoring his father insured he received as well as his own program of self-education, Boyle was thoroughly educated in a wide range of subjects—from languages (he was fluent in Latin, Greek, Italian, and French, and could at least read Hebrew and Arabic by the end of his life) to physics, mathematics, philosophy, and biblical criticism.

After two extended visits to what he referred to as the "illiterate nation" of Ireland in the early 1650s, Robert Boyle eventually settled in Oxford, in 1655, where he would remain until he relocated to London in 1668. His sister, Katherine, had arranged rooms for him in Oxford, and he continued to read intensely but also began to perform original experiments. He combined what he took from Hartlib with his reading of Pierre Gassendi (1592–1655), a French priest and mathematician who was a committed atomist, Francis Bacon (1561–1621), and Galileo, in order to undertake a course of completely overturning Aristotelian approaches to the natural world and natural philosophy. Casting aside such Aristotelian notions as forms that gave purpose to matter and natural place, Boyle instead focused on "corpuscularianism." Renee Descartes (1596–1650), the French philosopher and mathematician had first proposed this form of atomism, which posited that all matter is made of minute corpuscles, which possess shape, size, and motion. Thus, all material elements were describable in terms of corpuscles and their collisions, and chemical processes could be described in terms of the mechanical interactions of corpuscles. The power of this corpuscular theory was that it led to testable predictions about elements, as opposed to the Aristotelian theory of elements and principles. Thus, in spite of never giving up on alchemical notions such as sympathies and antipathies (the idea that certain elements are drawn toward other elements through action at a distance by inherent sympathies, or repelled by a similar process of antipathy), Boyle more or less single handedly founded the modern science of chemistry, through his meticulous application of the experimental method that was part and parcel of Galileo's mechanical philosophy, and his insistence on experimental testability over rationalist arguments as forms of proof.

However, while Boyle's approach to the study of nature was almost the opposite of that used by medieval intellectuals, the drive that motivated him was identical—he wanted to understand God better, by understanding what Boyle had no doubt was his creation, the world. In his 1659 work, *Some Motives and Incentives to the Love of God*, Boyle wrote:

> when with excellent Microscopes I discern in otherwise
> invisible Objects the Inimitable Subtlety of Nature's
> Curious Workmanship; And when, in a word, by the help
> of Anatomicall Knives, and the light of Chymicall Furnaces,
> I study the Book of Nature . . . I find my self oftentimes
> reduc'd to exclaim with the Psalmist, How manifold are thy
> works, O Lord? In wisdom hast thou made them all.

It was that wonder which drove Boyle to study and seek to understand the natural world, using the latest scientific instruments—such as the microscope mentioned above, which had been invented in 1590—and cutting-edge experimental methods to gain an ever-deeper understanding of what Boyle never doubted was God's creation. Late in life Boyle did acknowledge that some people might be put off from attempts to understand nature not only by the immensity of the task, but also by fears that it might promote impiety. Writing in 1686, in his *A Free Enquiry into the Vulgarly Received Notion of Nature Made in an Essay, Address'd to a Friend*, Boyle noted that

> the veneration, wherewith Men are imbued for what they
> call Nature, has been a discouraging impediment to the

Empire of Man over the inferior Creatures of God. For many
have not only look'd upon it, as an impossible thing to
compass, but as something impious to attempt.

However, in 1663 he had written: "If the omniscient author
of nature knew that the study of his works tends to make men
disbelieve his Being or Attributes, he would not have given
them so many invitations to study and contemplate Nature," in
his short work, *Some Considerations Touching the Usefulness
of Experimental Philosophy*. Although a modern reader might
take that to mean that the study of nature does, in fact, "tend
to make men disbelieve" in God, that was the opposite of what
Boyle actually meant, as is clear from the larger context of his
writings. Contained in the seven volumes of Boyle's surviving correspondence, forty-six volumes of papers, and eighteen volumes
of notebooks, are the notes and an outline on a never-completed
"little Tract about Atheism." In his many refutations of such lack
of belief, Boyle felt the strongest were the arguments from design,
which were apparent in "The Fabrick & Conservation of the world,
especially of Animals," as he put it. Boyle believed it was obvious
that a world with logically ordered, discernible and regular laws
of nature, was the product of a rational God. Thus, studying the
mechanical nature of God's creation could only strengthen one's
faith, rather than call it into question, and anything found in
nature that seemed to contradict God's divine power meant that
a misunderstanding had occurred. Just as Albert the Great and
Thomas Aquinas had argued centuries earlier, for Boyle there
could be no conflict between faith and reason.

In the 1650s Boyle was a member of a group of natural philosophers, which he termed the Invisible College, who exchanged
letters, provided moral support for one another, and in other
ways promoted rational inquiry. During the period following the
execution of King Charles I, Oliver Cromwell (1599–1658), who
had risen to prominence as a general fighting against the king
during the Civil Wars, ruled the country as a king in all but name,
and was succeeded by his far-less-capable and less-respected
son, Robert (1626–1712). After only a year, Robert was forced to
give up control of the country and flee into exile in France, and

would remain in exile for most of his life. The Invisible College was the spiritual precursor of the Royal Society of London, founded in 1660 when Parliament invited the exiled King Charles II (r. 1660–1685). Boyle was one of the founding members of the Royal Society, which would become the most consequential organization devoted to scientific research in the world, for reasons that we will soon see.

As a member of the Royal Society Boyle championed experimentation over the pure rationalism that Aristotelians favored to such an extent that his books now seem rather boring because he details the experiments he used in exhausting detail. However, it's only because such experimentation seems so obvious to those of us who grew up within an educational system in which experimentation is the backbone of scientific inquiry that his works drag on for the modern reader; for the seventeenth century they were completely revolutionary, and though Boyle never gave up on alchemical principles, his 1661 work *The Skeptical Chymist* really invented the science of chemistry. However, he also never lost interest in religion or changed his mind that the study of nature was an act of seeking to understand God through His creation, as evidence by such works as *The Excellency of Theology, Compar'd with Natural Philosophy* (1674) and *The Christian Virtuoso* (1690). That was the last work he published, just a year before his death, and it emphasized how understanding God as a clockmaker—as having imbued the physical universe with natural laws, which can be uncovered through the interrogations of the experimenter—served to reinforce one's faith. Furthermore, during these years he spent considerable sums to have the Bible translated into a variety of languages (including the first complete translation into Irish) in order to further missionary activities, and on his death he left a fund to pay for regular lectures to be given on the compatibility of religion and science. These Boyle Lectures were given at least annually for over 200 years, until they began to lapse in frequency in the twentieth century, but were revived in 2004.

For those interested in the historical relationship between religion and science, Boyle has been an interesting figure. White, in his *History of the Warfare of Science with Theology*, wrote that Boyle

> was at once bitterly attacked. In spite of his high position,
> his blameless life, his liberal gifts to charity and learning,
> the Oxford pulpit was especially severe against him, declar-
> ing that his researches were destroying religion and his
> experiments undermining the university. Public orators
> denounced him, the wits ridiculed him, and his associates
> in the peerage were indignant that he should condescend to
> pursuits so unworthy. But Boyle pressed on.

Draper, on the other hand, wisely passed over Boyle completely, except for a single reference to "some of the greatest scientific triumphs . . .its attempts to bring chemistry under the . . . laws . . . of Boyle." That's because Draper is the better—really the only—historian between the two, for all his flaws, and he had a better understanding of the controversies that surrounded Boyle, which had everything to do with his scientific approach and nothing to do with religious opposition to the work of this very devout man. As brilliantly discussed by Steven Shapin and Simon Schaffer in their book, *Leviathan and the Air-Pump: Hobbes, Boyle, and the Experimental Life,* traditional academics such as Thomas Hobbes (1588–1679) reacted negatively to Boyle's experiments not out of religious belief, but due to disagreements over his methodology. Hobbes represented a tradition that defined philosophy as certain and man-made knowledge, arrived at through rational analysis, and he recognized rightly enough that not only could Boyle's experiments never rise to the level of true certainty, but many of his experiments produced results that couldn't be reproduced (and thus, even by modern scientific standards wouldn't count as scientific). Boyle, on the other hand, admitted that his experiments could only lead to probabilistic truth—results that were statistically likely to be true, but could never be said to be absolutely certain to be true.

As it turns out, that latter standard of probabilistic truth was the one adopted as the standard by the Royal Society of London, no doubt in part because Boyle was a founding member. While today we see this as the best form of knowledge production in science (and many would argue it's the best form of knowledge production, full stop), it was a reversal of traditions going back to Aristotle and Plato, so there is no reason why it should have simply naturally won out. But this more modern approach to research and the production of knowledge, in which scientists were "modest priests of nature," as the members of the Royal Society sometimes referred to themselves, was well suited to the political and social circumstances of seventeenth-century England. This was a time in which the English elite classes looked back on horror at the ravages of the Civil Wars, which were in part the result of sectarian religious divisions, and longed for nothing more than a peaceful civil society that would allow them to carry on with the important business of making money. Thus, natural philosophers who modestly accepted the results of experimental investigations were very different than the religious enthusiasts—such as Oliver Cromwell—who had torn England apart during the Civil Wars. Therefore, the Royal Society's approach of only publishing natural philosophical findings that included a methods section detailing an experimental approach and probabilistically true results became standardized as the only acceptable approach to the study of the natural world, since gentlemen flocked to the prestigious and royally supported society. And since the Royal Society was the primary publishing venue for natural philosophy in England—what was emerging as "science" in the modern sense of the term—the Society's influence would spread across Europe and the British colonies as they were established around the world. Furthermore, in order to get published, it became increasingly necessary that one have the proper credentials—meaning a degree from a university—so the Royal Society played a tremendously influential role in standardizing not only what science was, but also what it was to be a scientist.

However, Boyle wasn't the most influential member of the Royal Society in its early days. That honor would go to his younger contemporary, and occasional correspondent, Isaac Newton (1642–1727). Newton's life was one in which historical chance repeatedly played significant roles, showing the importance of contingency in history, for there was no necessary reason why he should become the most dominant intellectual figure of not only England, but also all of Europe, when he was born the son of an illiterate farmer in 1642, during the turmoil of the Civil War. As Carl Sagan vividly described in *Cosmos,*

> Newton was born on Christmas Day, 1642, so tiny that, as his mother told him years later, he would have fit into a quart mug. Sickly, feeling abandoned by his parents, quarrelsome, unsociable, a virgin to the day he died, Isaac Newton was perhaps the greatest scientific genius who ever lived.

Neil deGrasse Tyson certainly agreed with Sagan, as demonstrated in episode 3 of his version of *Cosmos,* "When Knowledge Conquered Fear," where Tyson states that Newton's work challenged the idea that God had planned out the universe. However, though there were those who followed Newton who certainly saw such an implication in his work, as we'll see, Newton would have disagreed in the strongest possible terms with the idea that any of the ideas he developed challenged the notion of a providential God.

Newton is a figure who has stirred people's imaginations since at least the day he died, and he has attained a cult-like status among scientists that in all likelihood exceeds even that of Galileo. Inevitably, some have sought to paint him as somehow a martyr to science, as Galileo was. Andrew Dickson White seems to have started this trend, writing that "It was vigorously urged against him that by his statement of the law of gravitation he 'took from God that direct action on his works so constantly ascribed to him in Scripture and transferred it to material mechanism,' and that he 'substituted gravitation for Providence.'" But pay careful note to that passive voice construction—"it was vigorously argued." Who made that argument? White put the statement in quotes, making it appear that he was quoting some actual person, which is perhaps why others have quoted White again and again. M.B. Chapman, the editor of the *St. Louis Christian Dispatch* quoted from White in an 1894 issue of *The Methodist Review* in a way that made it appear the German mathematician and natural philosopher Gottfried Leibniz (1646–1716) may have made this statement, but that's not the case. Others try to clear things up with no more justification, such as Harry Fosdick in his 1922 book, *Christianity and Progress,* who ascribes this quote to an unnamed preacher. More recently, in his oddly named 2015 book, *Naturlopy,* Trung Nguyen writes that "Newton kept his true religious beliefs secret, for fear of persecution, literally until his dying day," and attributes the quote White provides (without mentioning White) to "Christians." While Newton was cagey about the specifics of his religious beliefs, it wasn't because he was in any doubt about the existence or importance of God, and the truth about his life is far more complicated than any of the writers (or the dozens, if not hundreds, of others who have used and reused White's "quote" without acknowledging where it derived from or that no one knows who, if anyone, White was actually quoting) would maintain.

As noted, chance was a very important factor in Newton's life, beginning with the death of his father, which occurred while Newton was still in the womb. His mother remarried a sixty-three year-old man when Isaac was not quite three, and he was left in the care of his maternal grandparents. His grandmother, Margery Ayscoug, unlike his father, was literate and prized reading, so she immediately began teaching her young grandson to read, assisted by his mother helped, who also encouraged her son in his intellectual pursuits when she visited. Thus, though unfortunate, the death of Newton's father actually had positive effects on the outcome of his life, as it's unlikely that Isaac would have learned to read if he had been raised in his father's home. Nevertheless, Isaac's separation from his mother was hard on him, and he clearly hated his stepfather—at the age of nineteen Newton compiled a list of sins he'd committed, which included "Threatening my father and mother . . . to burn them and the house over them." In 1653, after the death of her elderly husband and the birth of three more children, Newton's mother returned to their farm and Woolthorpe, and sent Isaac on to boarding school in Grantham—where her brother was headmaster—from 1655 to 1659. Isaac then returned to manage the family farm, and as it happens he was terrible at farm management. Therefore, his uncle, the headmaster of Grantham, convinced his sister to send Isaac on to Trinity College, Cambridge, in 1661, in spite of Isaac being somewhat older than most of the other freshmen (who tended to enter university around the age of fifteen at that time).

At first, Newton's education at Cambridge followed a traditional, classical approach, meaning that it was focused primarily on the works of Aristotle. However, by 1664 he began to move beyond this classical approach to education, reading the latest works of Continental philosophy, with a special interest in the writings of French philosopher, Renee Descartes (1596–1650). These works included such books as Descartes *Discourse on Method*, but also a Latin translation of Descartes' *Geometry*, which would have a profound impact on him. Soon Newton was delving into all the most important mathematical works available, from authors such as William Oughtred (1574–1660), an Anglican minister and mathematician who, among other achievements, invented the slide rule (the most important mechanical computing device before the invention of the electronic calculator), Francoise Viète (1540–1603), a French lawyer and mathematician who laid the foundations for modern algebra, and perhaps most important, John Wallis (1616–1703), a cryptographer, Presbyterian minister, and (from 1649) holder of the Savilian Chair of Geometry at Oxford University.

In 1665 historical contingency again reared its head, as Cambridge University temporarily closed down in response to what would turn out to be the last major outbreak of bubonic plague in Europe. As a result, Newton returned to the family farm for all but three months between 1665 and 1667, during which time he experienced what historians have termed his *annus mirabilis* (miraculous year). Newton made important experimental discoveries in optics, developed the mathematical theory of uniform circular motion—the Dutch astronomer and mathematician, Christiaan Huygens (1629–1695) had developed this theory in 1659, but didn't publish his findings until 1673—and perhaps most importantly, he wrote a tract in which he laid out the rules and methods for calculus in 1666, almost a decade before the

German polymath Gottfried Leibniz (1646–1716) independently developed calculus. Newton's development of what would turn out to be the most important mathematical tool for the study of physics was built in part on the work of John Wallis, and Newton's failure to publish his results contributed greatly to what would turn out to be a very rancorous disagreement with Leibniz, who never accepted that Newton had beat him to the development of calculus.

Newton returned to Trinity as a fellow in 1667, where he continued his work, but by this point he was already the leading mathematician in England. It was in recognition of this fact—and in response to Newton expanding his short tract on calculus—that Isaac Barrow (1630–1677) recommended Newton replace him as the Lucasian Professor of Mathematics, only four and a half years after Newton received his B.A. Over the next fifteen years, Newton wrote a book-length study of calculus that he completed in 1671, but failed to find a publisher for, continued his research on optics, lectured extensively on algebra, engaged in exchanges of letters with a wide-ranging group of European intellectuals (including Huygens and Leibniz) about such topics as calculus, and finally published his findings on optics in 1672 in the *Philosophical Transactions of the Royal Society*. This led to four years of acrimonious exchanges with scientists such as Robert Hooke (1635–1705) and Christiaan Huygens that were so upsetting to Newton that he largely withdrew from public scientific discussions. As a man of humble beginnings, Newton lacked the natural confidence of men such as Robert Boyle, and he took criticism very badly.

During this period, Newton didn't limit himself to what fit the modern definition of science, however. In fact, he was actively engaged in both alchemical experimentation and theological speculation. From an alchemical perspective, Newton wrote over a million words on the subject over the course of his lifetime and engaged in research so frequently that it seems to have contributed to the periodic mental breakdowns he experienced late in his life—chemical analysis of hair samples left behind by Newton shows a high concentration of the heavy metal mercury (often known as quicksilver), which is central to alchemical research. Mercury can easily be absorbed through the skin, and the symptoms of mercury poisoning include fatigue, depression, lethargy, irritability, and headaches—all of which Newton experienced at various times to such a degree that he was sometimes forced to withdraw into seclusion, so it appears that he went through cycles of alchemical research that exposed him to mercury, an apparent breakdown that would lead to his seclusion for a length of time that would allow his body to purge itself of enough of the mercury for him to recover, followed by a return to alchemical research and . . . well, you get the picture. We should be mindful of how engaged Newton was in alchemical research throughout his lifetime, and suspicious of modern claims that consideration of this work is pointless—like that made by Neil deGrasse Tyson, that "this work led nowhere." In fact, two important alchemical principles—that of sympathies, with like substances naturally attracting one another at a distance, and antipathies, that unlike substances naturally repel one another—would prove extremely important to Newton's work, as we'll see momentarily.

That Newton never attempted to publish any of his alchemical work shows the declining place this subject held in the European intellectual world, as chemistry slowly separated itself from alchemy—thanks in no small part to the work of the alchemist Robert Boyle. As modern people, we would like to think that alchemy became marginalized due to its lack of results, but in reality, its marginalization was largely the product of the success of the mechanical philosophy the Royal Society promoted, which denied the existence of such things as action at a distance. This became the dominant approach to natural philosophy in the seventeenth century—after all, one couldn't get published by the Royal Society without adhering to its tenets—so it was due to social forces rather than pure rationalism that the newly emerging class of gentlemen scientists came to sneer down their collective noses at alchemy, in spite of the occasional holdout such as Newton.

Fortunately, Newton didn't need a public acceptance of alchemy to become famous. In 1679, Robert Hooke, in his capacity as the secretary in charge of correspondence for the Royal Society, wrote Newton in an effort to engage him in the problem of planetary motion, in hopes of Newton producing something publishable for the *Philosophical Transactions of the Royal Society,* which is, it should be noted, the oldest science journal in the world. It seems that Hooke had little idea of how much his earlier criticism of Newton's work had angered the younger man—to Hooke, it as an issue of intellectual disagreement, but to Newton, it seemed intensely personal—and Newton refused to reply to Hooke directly. Nevertheless, he did write back in general terms (leading to further disagreement from Hooke) and the exchange stimulated Newton in very productive ways. He received further stimulus when a rather bright comet put on a show in the winter of 1680 and 1681, which set England's community of natural philosophers abuzz with conversation.

The problem was that Kepler had shown planetary orbits to be ellipses, but others—such as Renee Descartes—had made powerful arguments for the straight line as the natural pathway objects in motion would follow if not acted on by outside forces. And that qualifier of "outside forces" was extremely important to the emerging view of mechanical philosophy as the definitive approach to understanding the natural world. Planets move in ellipses, as do comets such as the one observed in 1680/81, so what outside force acted on these bodies to keep them from moving in a straight line? Robert Hooke posited to Newton that the natural tangential motion of bodies in motion combined with an attractive motion toward a body's center explained planetary (and cometary) motion, but he couldn't work the math out—or explain why the moon didn't fall to Earth if it was being drawn toward the center of the planet, or for that matter why all the planets didn't fall into the Sun. Hooke, like most scientists of his day—and yes, I realize I'm using the term "scientist" and "natural philosopher" rather interchangeably during this period of transition from natural philosophy to science—was committed to the idea that there had to be some mechanical force at work, in order to explain these apparent discrepancies between observed reality and philosophical notions about what should happen.

However, Newton worked from a very different set of principles than Hooke. In 1675, Newton had published a short paper titled "An Hypothesis Explaining the Properties of Light" in *The Philosophical Transactions of the Royal Society.* In this paper, he had stated that "God, who gave animals self-motion beyond our understanding, is, without doubt, able to implant other principles of motion in bodies." In other words, God acted on matter in ways that might not be obvious to mechanical philosophers, even if His actions were logically consistent and mathematically describable. And one of the many ways that he thought God acted in the universe was through "certain ethereal spirits" infusing nature, with properties that include sympathies between like and antipathies between unlike substances that any alchemist would recognize. It was precisely this manner of thinking that led Newton to envision the Moon as indeed moving in a straight line, but being pulled toward the Earth by a sympathy between lunar and terrestrial substances, while pushed away by antipathies between other substances in the two bodies. Alerted to the back and forth between Newton and Hooke, one of the best-known astronomers in England, Edmond Halley (1656–1727) visited Newton in Cambridge and asked to see his solution for the problem of planetary motion in 1684. Newton stated that he couldn't find his calculations, so he would redo them, resulting in a short work of only nine sheets, which he sent to *The Philosophical Transactions of the Royal Society* as "De motu corporum gyrum," or "On the Motion of Bodies in an Orbit."

Newton's short work contained two central insights: one, that the Moon *is* a falling body—it just never quite lands, as it were, on the Earth; two, that the question of what precisely is the mechanism causing this motion need not be addressed, as long as the math can be worked out. Unsurprisingly, this led other scientists to complain that Newton was appealing to occult forces and a belief in action at a distance because . . . well, he was. If it hadn't been for his religious and alchemical beliefs, he'd never have proposed this odd notion. Yet, the math worked out, doing a better job of describing celestial motion than anyone had managed yet. Furthermore, Newton's "On the Motion" had numerous general implications for the motion of bodies that begged for further work to be done. At Halley's urging—and with his financial support for publishing the result (Halley was the wealthy heir to a fortune built on soap manufacturing)—Newton expanded his work over the next three years into a massive work of just over 1,000 pages divided into three books, the *Philosophiæ Naturalis Principia Mathematica* (*Mathematical Principles of Natural Philosophy*, which I'll refer to as the *Principia* from here on out), published in 1687.

In the preface, Newton stated that "Rational Mechanics will be the science of motions resulting from any forces whatsoever, and of the forces required to produce any motions, accurately proposed and demonstrated" and that these forces could be understood through the application of mathematical principles, supplied in the *Principia*. Though often expressed through the language of geometry, Newton used the methods of the calculus he had invented in order to describe motion and offer up powerful predictive tools for understanding how objects interact with one another—even if he sidestepped the issue of what the force might be that causes objects to interact with one another at a distance. Newton used a Latin word for weight—*gravitas*—to describe this force, which is where we get our word "gravity." Many critics—such as Gottfried Leibniz, upset over the question of who invented calculus first—jumped on this point, but there was no denying the power of Newton's mathematical proofs. For example, Newton offered up a crucial proof about the rate of the Moon's "fall"— that is, its orbital velocity. Newton was able to show that its rate of fall/orbital velocity was exactly what one would expect given its distance from the Earth, by working out the mathematical relationship between the rate of fall of objects falling at sea level versus what one would expect given the Moon's distance from Earth.

The *Principia* might be the most important book ever written in the history of science. In many ways it was the culmination of developments that had begun with Copernicus, to be elaborated by Kepler, Brahe, and most importantly before Newton, Galileo. It ushered in what was considered the definitive approach to understanding physics—particularly when it came to celestial mechanics—until the time of Albert Einstein (1879–1955). True, it would engender critiques of religious thought in the eighteenth and nineteenth century, but that was certainly not Newton's intent. This was the work of a highly devout man, who wrote in the *Principia* that "this most beautiful system of the sun, planets and comets, could only proceed from the counsel and dominion of an intelligent and powerful Being." Some, such as the tremendously important historian of science Richard S. Westfall (1924–1996), have argued that such statements can be explained by portraying Newton as a deist—one who believes in a god who created the universe, then stepped back from it and let natural law have its way after that initial act of creation—but that view doesn't mesh well with the evidence.

For one, Newton engaged in vigorous intellectual debates from a religious standpoint, such as the long-running arguments he had with Leibniz. Newton insisted that absolute space is a thing—that the universe is finite in size and that space itself has positive existence—versus Leibniz's insistence that space is just the nothingness between planets and stars. Newton's basis for his position was that God had created the universe from a point outside of it, and thus space had to have an existence of its own. How else could God be outside of space, if space lacked being? Such a position could conceivably be held by a deist—but Newton's position on the repeated elliptical orbits of planets is harder to square with the idea that he was a deist. Beginning in 1710 Leibniz began to openly attack Newton's concept of gravity as a form of miraculous intervention, and in 1715 Leibniz attacked Newton for positing a universe in which God "had not . . . sufficient foresight to make it a perpetual motion." This was because Newton had noted that repeated elliptical orbits were never exactly the same, and that without some correcting mechanism, it appeared that planetary orbits would eventually collapse into chaos. Newton suggested as a solution that God caused comets to strike the sun on occasion, to power it up as it were and thereby to keep planetary orbits in line.

But more important as a refutation of the idea that Newton was a deist is what he didn't publish. Just as he wrote more than a million words on alchemy and performed numerous alchemical experiments, Newton expended a great deal of time and energy writing about religious topics, with prophecy being a favored topic. He was particularly fascinated with the book of Daniel, and developed numerous complex formulae in an effort to predict the future by using the bible. He was unorthodox in his beliefs—he rejected Jesus' divinity, for example, and he also rejected the idea that demons had real existence—but it's obvious from all the work he put into his religious writings that these were of great importance to him. And though he did think classical sources could supplement and sometimes be more reliable from a historical standpoint than the bible (which is why Westfall argued that Newton was a deist), he never rejected the authority of the bible when it came to divine subjects. Even as a historical source, Newton tended to defer to the bible, as is clear from a letter he wrote to the English theologian Thomas Burnet (c. 1635–1715) in 1681. Burnet was about to publish his book, *A Sacred Theory of the Earth*, in which he provides a philosophical and rational interpretation of Genesis' story of creation. Newton (like Burnet, and most other theologians of his time), stated that so long as one interprets the Genesis account allegorically, there can be no better description for the formation of the Earth. This is in keeping with the "postulates" of William Whiston (1667–1757), an Anglican priest and mathematician who was an ardent popularizer of Newton's works—and whose career Newton supported in turn. Whiston wrote that

I.   The obvious or literal sense of Scriptures is the true and real one, where no evident reason can be given to the contrary.

II. That which is clearly accountable in a natural way, is not without reason to be ascribed to a miraculous power.

III. What ancient tradition asserts of the constitution of nature, or of the world, is to be allowed for true, where 'tis fully agreeable with Scripture, reason, and philosophy.

Whiston was a talented mathematician who succeeded Newton to the Lucasian Chair in Mathematics at Cambridge in 1702, and like Newton rejected Trinitarian Christianity (which led to his expulsion from Cambridge in 1711). There were tensions between the two—particularly over differences in biblical prophecies—but as revealed in Newton's surviving letters and religious writings, he was in agreement with Whiston over how to approach the bible. Thus, in spite of what Westfall believed, Newton did see the bible as a divinely inspired work that should be given great reverence—he just felt that it sometimes needed allegorical interpretation and that it wasn't infallible as a work of history.

Newton found himself embroiled in controversy on repeated instances, and at times found himself the subject of suspicion due to his religious beliefs. However, this was decidedly not because he was a deist or atheist, and he was not attacked because of his scientific work. Instead, many felt (rightly, as it would turn out) that he wasn't orthodox in his religious beliefs. As a modern person, I'm completely on board with the idea that he should have been left to his Unitarian beliefs without facing discrimination about them. But the point is, contemporaries never attacked him on the grounds his scientific work challenged religious views—the religious suspicions Newton aroused were based entirely on his doctrinal views and attitude toward scriptural interpretation. And even those suspicions were muted. In the 1690s he published a number of works on scriptural interpretation, which were widely respected. Furthermore, far from facing discrimination, after the publication of the *Principia*, he bounded from success to success: in 1689–1690 and 1701–1702 he sat as a Member of Parliament for Cambridge University, in 1699 he took up the prestigious (and lucrative) post of Master of the Royal Mint, and in 1703 he was made president of the Royal Society and an associate of the French Académie des Sciences (more about this organization shortly). In that role, he held a virtual stranglehold over what the Society would or would not publish, and who it admitted into its ranks. This had a tremendous impact—as far as publication went, it meant that only those who adhered to a probabilistic view of truth supported by empirical data analyzed with mathematical rigor could hope to get published, and as far as membership in Royal Society went, holding an appropriate university degree was the minimum qualification. It was very difficult for non-gentlemen to get into Oxford or Cambridge, and those who did tended to get assimilated into the class of gentlemen by virtue of acquiring Oxbridge manners, manners of speech, and attitudes. Culture is a powerful thing, and the gentlemen who made up the student body at these universities tended to heap abuse upon those who didn't dress, act, or behave (or for that matter think) like a gentlemen. Thus, the ideal of the gentleman scientist firmly took root under Newton's presidency. Finally, in 1705 Queen Anne (r. 1702–1714) knighted Newton during a visit to Cambridge.

At first, Newton's influence was largely confined to the newly United Kingdom (England, Scotland, and Wales were officially joined as the United Kingdom in 1707). Scholars, like everyone else, tended to be nationalists. The Germans had their own geniuses in men such as Leibniz, who died in 1716 still hating Newton and fighting against everything he represented. The French tended to follow the natural philosophy of Rene Descartes, which had much to offer—but was both rationalist and utterly mechanistic in its approach. In other words, natural philosophers working in the Cartesian tradition privileged thinking through problems and creating philosophically sound arguments for their solutions instead of using an experimental method they tended to view as lacking the surety of logical reasoning. As for their mechanistic leanings, like Galileo they rejected the concept of action at a distance and believed reasonably enough that any observed effect had to have a mechanical cause. However, the Continental reception of Newton and Newtonianism would change in the eighteenth century due in no small part to François-Marie d'Arouet (1694–1778), who is far better known by his chosen pen name, Voltaire. He is one of the more important non-scientists in the history of science.

Although he would not adopt the name until 1718, we'll use the name "Voltaire" in order to avoid confusion. Voltaire was born into the France of Louis XIV (r. 1643–1715), the most absolute monarch in Europe. France had no body comparable to the English Parliament, and in a very real sense, Louis' word was law.

French monarchs had a tendency to hire men of talent from the middle class as civil servants and pay them poorly, but allow them abundant opportunities to become wealthy through the solicitation of gifts or outright graft, and such men were often ennobled. It was a good time to be a bureaucrat in France, which is what Voltaire's father was. His mother was an aristocrat, and Voltaire was the fourth of five children. Given this background, Voltaire received a top-notch education from the Jesuits at the Collège Louis-le-Grand in Paris, and his father—who had extensive literary contacts—encouraged his son to read broadly. However, the elder François was also determined that his son would go on to law school and one day become a bureaucrat, thus becoming wealthy through the pursuit of graft—just as François had done.

Voltaire, however, had other plans. This was the age of the Enlightenment, a period in which critical thinking and self-reflection was privileged, when ancient sureties—such as the idea that the monarch held his position by divine right, as God's representative on Earth, was not only open to question, but questioning was encouraged, at least in intellectual circles. Even the status of the Church—the status and even existence of God himself—was wide open to question. It didn't slow the tide of critiques and questions that France was a decidedly unjust society, with an autocratic monarch backed by a church staffed by men who were often more devoted to living well than caring for their flock. Paris was the epicenter of the Enlightenment, and it was rife with literary salons—places where wealthy patrons would periodically have gatherings of philosophers, artists, and poets, where the ideas would flow as freely as the wine—coffee houses, book shops, and other venues of free thought. This was the Paris where Voltaire's father sent him to work as a notary. However, this work took second place to Voltaire's primary past time of ingratiating himself with the Parisian intellectuals, writers, and ne'er do wells (all those categories tend to blur together anyway) with his wit, quick tongue, and the ability to write works in either verse or prose that tended to blend intellectualism with humor. Eventually his father learned of his freewheeling, writerly lifestyle, and sent him away to Caen to study law. Continuing to spend more time at parties and writing plays than at study, in 1713 his frustrated father acquired a post for him as secretary to the ambassador to The Hague in the Netherlands. As a secretary, Voltaire was a failure, but as a lover, he had real talent. He fell in love with a French Protestant refugee with a better nickname than any historian could ever hope for—Pimpette—an affair that was discovered before the end of the year, leading Voltaire's panicked father to bring him home.

In order to understand Voltaire, his views, and what would become of him, we should pause a moment to consider what France in the late seventeenth and early eighteenth century was like. As mentioned, the king was effectively an autocrat, ruling without any real concern for what the people wanted or needed. Louis XIV had a massive army that engaged in frequent wars and had spent enormous sums building a palace at Versailles, just outside Paris, that is still breathtaking to this day. Society was divided into three classes known as "estates." The First Estate was made up of the clergy, who enjoyed tremendous privileges. The Catholic Church (keep in mind that the French king was known as "the most Catholic king," and was very closely allied with the Catholic Church) owned about 6% of the land in France and all French farmers were legally obligated to hand over 1/10th of their crops (or a monetary equivalent) per year. Although there were a great many poor monks and priests who were motivated by their love of God, there were also plenty of indolent members of the clergy, monks who ignored the dictates of the rule they were supposed to live under, and upper-level churchmen who were normally from the aristocratic class and often couldn't care less about their flock or their religious responsibilities.

The Second Estate was certainly no better. Making up some 1.5% of the population, the nobility who made up this class controlled between 20 and 30% of the land in France—yet astoundingly paid no taxes. The argument was that nobles provided service to the king as advisers or in time of war, but in reality this was a blatant bribe to keep them from revolting against the autocratic king. The Third Estate, then, was made up of everyone else, lumping teachers, lawyers, merchants, bankers, workers, farmers, and the poor into one overtaxed mass that had little or no voice in either local or national government. It's small wonder that the best future people such as François d'Arouet—Voltaire's father—could imagine was one that allowed them to become as rich as possible through the opportunities for corruption available to bureaucrats while dreaming of becoming ennobled, thereby exchanging the crushing tax burden imposed by the king for the endless rounds of parties, salons (intellectual gatherings), and dances available to the nobility.

It was within this environment that Voltaire matured into a committed social critic, who rejected much of the established dogma of what was acceptable in France, in large part because he felt there was something decidedly rotten about French society. This would cause him to be very popular, as he made his critiques earthy and humorous, but could also get him into considerable trouble. In 1717 French authorities arrested him for writing a satirical poem critiquing the government of the regent, Philippe II, Duke of Orléans (1674–1723), who ruled France in place of the five-year-old Louis XV (r. 1715–1774) from the time of Louis XIV's death in 1715 until his own death in 1723. The Regent needed no excuse to arrest Voltaire and the criticism of his free-spending and scandal ridden government alone might have been enough to have landed the writer in jail, but if not, the accusation that Phillippe had committed incest with his own daughter was quite enough.

Voltaire spent eleven months in jail and emerged to become the darling of French society with the publication and production of his play, Œdipe (both the Regent of France and the king of England awarded Voltaire with medals for this play). Coupled with financial support from various wealthy patrons and his own successful investments, Voltaire could now live a life of the Parisian writer and intellectual—at least until he found himself in legal trouble for a second time. In 1726 a powerful young nobleman, Guy Auguste de Duc de Rohan (1683 –1760), taunted Voltaire about his recently adopted name. Voltaire responded, quite correctly as it turned out, that his own name would be honored forever, while the Duc de Rohan would bring shame to his own name. Infuriated, the young nobleman sent some of his retainers to beat Voltaire, horsewhipping him and then grinding salt into the wounds. Voltaire responded in typically French fashion by challenging the Duc de Rohan to a duel, but the young man's family instead arranged to have an arrest order put out for Voltaire. Fearing that he would spend the rest of his life in the Bastille this time, Voltaire instead fled to England and would remain in exile—spending most of his time in England— until 1729.

In England, Voltaire found a society very different from his own. It would be wrong to exaggerate how free English society really was, but in comparison to France it was a libertarian paradise. England had a king and one could certainly get into trouble for being overly critical of him in print, but more often through the exercise of informal social pressure rather than resorting to legal means. Pamphlets were printed, cartoons were drawn, and speeches were made against him in Parliament and no one got horsewhipped or arrested. And speaking of Parliament, not only did it have real power—England was well and truly a Constitutional Monarchy by this period, with the memory of Charles I's trial and execution serving to remind any monarch who got too uppity about how important Parliament really was—but there were vying political parties. Elections weren't democratic by any means—typically a group of county elites would decide who was going to be the Member of Parliament for any given county before any actual voting was done—but the House of Commons was filled with merchants, bankers, lawyers, and other members of the well-to-do but non-noble class, giving a political voice to people who were voiceless in France. From time to time, MPs even expressed a small measure of concern for the poor. England's official religion was the Church of England, and Catholics as well as a growing number of "dissenters"—Protestants who weren't Church of England—suffered from a variety of political disadvantages. But the days of hunting down and imprisoning or killing religious minorities was long past in England. And as for the Church of England, its clergy tended to be rather well educated and well intentioned, demonstrating real concern for their parishioners (for the most part, there were of course bad apples) and attempting to do their jobs to the best of their ability.

Voltaire fell in love with England and, in many ways, English ways of doing things. He was particularly impressed with Sir Isaac Newton's funeral. This was a man who had been born into modest circumstances and had risen to the peak of English society by virtue of his merit. Newton's funeral was conducted as an affair of state, and he was buried with full honors in Westminster Abbey. It's still possible to see his grave and the impressive monument that marks it, and it's certainly worth a visit. It's right next to the burial site of the astronomer Sir William Herschel (1738–1822), whose discovery of Uranus offered a powerful proof of Newton's theories. Seeing Newton's elaborate funeral left a lasting mark on Voltaire.

Voltaire would eventually travel back to France in 1729. He made a small fortune by buying into a lottery the French government held in order to pay some of its staggering debt, and laid claim to another fortune from his father's inheritance, allowing him to write what he pleased and how he pleased, within the limits of the French legal (and social) system. Much of what he would write would be couched in satire in order to avoid further complications—or at least that was the hope. He would again be forced into exile in 1734 after publication of his most important work, the satirical *Persian Letters*. While this work is important for many reasons, for our present purposes its greatest importance is that it lampooned an enemy that existed more in Voltaire's imagination than in reality (though not *that* much more), French Academic Cartesianism.

The problem was partly one born of nationalism, and partly one born out of the structure of what was now termed the French Royal Academy of Sciences, after the Academy of Sciences received a royal charter in 1699. The English Royal Society could be overly rigid and sometimes driven by personal prejudices (Newton disallowed publication of a rival's work on at least one occasion when he was its president) but in many ways it was rather open. Membership was (and is) open only to those who are invited to apply, but the board making such invitations was rather large and diverse, and while members received certain privileges, the most important one to be gained was the social prestige associated with being able to append an F.R.S. (for "Friend of the Royal

Society") to one's name. Certainly there were standards meant to keep out those not deemed gentlemen (including a requirement to pay dues) but membership was fairly widespread. None of this was true of the French Royal Academy of Sciences. Louis XIV's financial controller had founded the Academy of Sciences as an organ of government in 1666, and that association with government was only strengthened when the Academy received a royal charter in 1699. The king appointed the Academy's president, and membership was hierarchical. The most sought-after positions within the Academy were, naturally enough, those of the pensioners—the senior members who received a royal pension—with associates and assistants falling further down on the hierarchy.

This structure meant that it was far easier for the king to exercise his authority over the Academy, and its members tended to be even more conservative and resistant to new ideas than their counterparts in the British Royal Society. Thus, the upper echelon of the Academy championed Cartesianism—which was suspicious of (though it did not completely reject) experimental science and probabilistic models of truth—and most importantly for the sake of Newtonianism, rejected action at a distance. Cartesianism posited the existence of monads that infused physical objects, pushing and pulling them this way and that—and from the standpoint of observation, this model was actually quite powerful. However, younger members of the Academy were increasingly dissatisfied with Cartesianism, in part because they found Newtonian physics to have greater predictive power—but frankly, in large part because they were tired of being shut out of the best positions in the Academy (and publication!) by men they viewed as old fashioned and out of touch.

While we would like to think that scientific progress occurs as better ideas win out, in reality that's only partially true. As the work of scholars such as Thomas Kuhn (1922–1996) and Pierre Bourdieu (1930–2002), among a great many others, has made clear, a variety of social, cultural, and economic factors play very large roles. The prestige of those holding any given idea as true, cultural attitudes and attachments, and money are all tremendously influential when it comes to the contest of ideas. That's why Voltaire is so important to what scholars such as J.B. Shank have termed "The Newton Wars." Although Voltaire would have avoided trouble over his *Philosophical Letters* if he had been able to do so, the fact that he was forced yet again to flee into exile made him appear to be a rebel who was willing to risk life and liberty in order to stand up for what he thought was right. He achieved something of the status of a rock star in early eighteenth-century France. He had begun a love affair with a married mother of two who was twelve years younger than him— and brilliant in her own right—Emilie, the Marquise Du Châtelet (1706–1749) and it was at her husband's home in Cirey—which is in northeastern France, but was legally beyond the reach of Voltaire's enemies—where he took refuge (you really have to love the French . . .).

The fact that Voltaire was living in exile with a younger woman did nothing to detract from the esteem Voltaire's fans held him in. And the fact of her brilliance would help the cause of Newtonianism tremendously. Fluent in multiple languages and a skilled and talented mathematician, Emilie translated the *Principia* from its turgid Latin into a lively and accessible (for a work of advanced physics, that is) French, making Newton's ideas more accessible even as Voltaire popularized Newton's ideas and undercut the reputation of Cartesians, whom he portrayed as stodgy old fools. The most important contribution in this vein was Voltaire's 1738 *Elements of the Philosophy of Newton,* which showed clear signs of Emilie's influence. She published her own *Foundations of Physics* in 1740, which was in no way meant for a popular audience—it instead showed her as the intensely brilliant mathematician that she was.

In sum, by the 1740s the tide had begun to turn against Cartesianism, and by 1750 this school of thought was seen as decidedly old fashioned. Much of the reason why was because of the reputation, talent, and name recognition of Voltaire and his lover, Emilie, the Marquise Du Châtelet. Voltaire found her company delightful—and as an aside, her husband never had any concerns about their relationship—but eventually found Chirey stifling, and left for Holland and Prussia. Emilie found another lover—a poet ten years her junior—in 1748, and became pregnant with his child, dying in childbirth in 1749. Voltaire continued to bounce around Europe and took a number of lovers—including his own niece, who would inherit most of his property at his death—and wrote prolifically. He spent most of the latter part of his life in Geneva, though he was able to return to Paris in the last years of his life—where, coincidentally he became friends with the American Benjamin Franklin (1706–1790). He died in 1778 and had to be buried in secret, since he was denied burial in consecrated ground due to his religious beliefs (his brain and heart were embalmed separately). In 1791 he was disinterred and reburied with great pomp and circumstance—his funeral procession drew as many as a million people—in the Pantheon, which was originally a church before being turned into a national mausoleum to house the great and well known in France.

From the standpoint of the history of science, Voltaire's position is important. His championing of Newtonianism and his lampooning of Cartesianism greatly contributed to the victory of the former over the latter among French intellectuals. And with the imprimatur of the French Royal Academy added to that of the English Royal Society, Newtonianism rapidly became seen as virtually the only permissible approach to physics (and in many ways, to knowledge production in general) across Europe. However, from the standpoint of the history of the relationship between religion and science in the West, Voltaire is even more important. As a committed deist—one who believes that God created the universe, complete with natural laws that keep it functioning, and then backed away, to intervene no more—he believed he'd found the perfect answer to how to understand a mechanistic universe devoid of divine intervention. Deism had begun to gain ground among certain groups of intellectuals in Europe following a massive earthquake that killed tens of thousands hit Lisbon, Portugal—a city dotted with religious institutions—had caused many to ask how a loving and omnipotent God could have allowed such a thing to happen. With Voltaire's championing of Newtonianism as an adjunct to deism, many began to see the relationship as a natural one. However, we must keep in mind that Voltaire arrived at this position due to a particular set of social and political circumstances, as a man who lived in a country with an oppressive political system supported by a deeply corrupt church. Perhaps he would have come to reject traditional Christianity if he had been born in England, for example—but perhaps not. Regardless, in spite of the naturalness of the fit between Newtonian physics and deism that Voltaire saw (or thought he saw), Newton would have vociferously disagreed with Voltaire's religious views. A deeply devout man (though an unconventional one) who believed in a God that was not only a creator, but actively involved in His creation, Newton would have been horrified by the use to which Voltaire and other deists put his physics.

Further Reading

Davidson, Ian. *Voltaire: A Life*. New York: Pegasus Books, 2012.

Force, James E. *William Whiston: Honest Newtonian*. Cambridge: Cambridge University Press, 1985.

———. and Richard H. Popkin. *Essays on the Context, Nature, and Influence of Isaac Newton's Theology*. Kluwer Academic Publishers: Dordrecht, 1990.

Hagengruber, Ruth, ed. *Emilie du Châtelet between Leibniz and Newton*. Springer: New York: 2012.

Hall, A. Rupert. *Isaac Newton: Adventurer in Thought*. Cambridge: Cambridge University Press, 1992.

Holton, Gerald and Stephan G. Brusch, eds. *Physics, the Human Adventure: From Copernicus to Einstein and Beyond*. New Brunswick: Rutgers University Press, 2004.

Shank, J.B. *The Newton Wars and the Beginning of the French Enlightenment*. Chicago: The University of Chicago Press, 2008.

Shapin, Steven and Simon Schaffer. *Leviathan and the Air-Pump: Hobbes, Boyle, and the Experimental Life*. Princeton: Princeton University Press, 1985.

Wade, Ira O. *Intellectual Development of Voltaire*. Princeton: Princeton University Press, 1969.

Zinsser, Judith P. *Emilie Du Chatelet: Daring Genius of the Enlightenment*. New York: Penguin Books, 2007.

CHAPTER 6

Revolution, the Changing Conceptions
of Creation and Creatures, and the
Rise of Fundamentalism

A s Europe passed through the eighteenth and into the nineteenth century, the continent experienced tremendous turmoil. It was barely more than a decade after the death of Voltaire that the problems in France reached a point of crisis. The tremendous debt, extreme inequality, and the stifling autocratic government that King Louis XIV had perfected were simply unsustainable, and in 1789 the French Revolution broke out. The near-term precipitating cause was the debt crisis of 1788. In the previous century, the government had lost several wars and the country had been the victim of free-spending kings; Louis XV, who reigned from 1715 to 1774, maintained an even more lavish court—and spent more on mistresses—than his great-grandfather, Louis XIV, whom he succeeded, while also fighting and losing global wars against the English, from the subcontinent of India to North America. Thus, by 1788 over half of all income the French government took in went just to paying the interest on the 1.6 billion *livres* the government owed. And since French kings had defaulted more than once on debts, no one was left who was willing to loan the government any money. In desperate efforts to stave off the inevitable the government had imposed ever-heavier taxes on its people—which fell almost entirely on the common people—which made their already difficult lives nearly impossible. Nearly, that is, until rains and crop failures made the cost of bread skyrocket, tipping things toward the impossible for the lives of the majority of people in Paris in 1789.

Meanwhile, in comparison to France, the British were fast becoming a world power during this same span of time. During the seventeenth century the British formed their own stock market—by copying the Dutch—and developed a national bank that allowed for far better management of the economy than the French could pull off. Furthermore, the Industrial Revolution was in full-swing by 1750, allowing British industry to be far more productive than their French manufacturing counterparts—who focused on small-scale, high-quality luxury goods. This allowed both for a robust tax base for the British government as well as a rate of production of weapons and munitions (among other things) never before seen in world history. As a direct result, the British were able to conquer areas such as the enormously wealthy Indian subcontinent, which came more or less fully under British control after the British victory at the Battle of Plassey in 1757. Therefore, the British were able to rob the people in their imperial possessions in order to bolster their own wealth, at a rate the French would never come to match.

As a result of the economic conditions in France, the people demanded change. These demands were at first moderate, but soon became increasingly radical in nature. Alarmed by events in France, European powers such as Prussia made threats and expressed support for the French monarchy, and by 1792 the infant French republic was at war with an increasingly large portion of Europe. Under pressure from within and without, the French revolutionaries became increasingly radicalized, and by 1793 the Convention (the elected government of France) had not only executed the king and queen, but had rapidly started executing all nobles and anyone who might express a hint of disloyalty to the government. Hundreds of thousands were arrested and tens of thousands were executed or allowed to die from poor treatment in prison. The Convention was first replaced by the Directory, then by the successful general Napoleon Bonaparte (1769–1821), who seemed determined to conquer all of Europe, until he was stopped by the Russian winter in 1812 and a grand coalition of Austria, Britain, Prussia, and Russia in 1814—and by Britain and Prussia again in 1815 when Napoleon briefly returned to power.

As the French armies spread across Europe, they took with them the ideas of the Enlightenment—including the deep suspicion of religion that many French intellectuals had. There's no reason to imagine that the average French soldier harbored such beliefs, but many of the officers did, and as more and more of Europe fell under control of the French, an increasing number of European intellectuals expressed non-traditional religious views such as deism or outright atheism. However, we need to be careful about what is meant by "an increasing number." There is no evidence that atheism, deism, or anything other than traditional religious beliefs became common among European intellectuals— much less among the common people. It's easy to mistake the outspoken viewpoints of well-known writers as somehow the norm.

Given the cataclysmic events of the European-wide wars of the late eighteenth and early nineteenth century, it's not surprising that scientific development radically slowed across much of Europe. However, it never came to a complete halt, even in France, in spite of the impact on science and scientists of the disastrous events that befell Europe. For example, Antoine Lavoisier (1743–1794) was a French nobleman who sought to apply the sort of mathematical precision found in Newtonian physics to chemistry, and in the process he was almost singlehandedly responsible for transforming chemistry into the quantitative (relying on close measurement and weighing, as opposed to descriptions of the qualities of a matter) science that it is today. In the process he overthrew what was known as phlogiston theory, which posited that a substance known as phlogiston existed in flammable objects. According to this commonly held view, the consumption of phlogiston allowed fire to burn. Thus, combustion was a process of subtraction, as fire consumed phlogiston. Lavoisier showed that oxygen—discovered by the Englishman Joseph

Priestley (1733–1804), though he mistakenly thought of it as air with all the phlogiston removed from it—combined with matter to cause combustion. Thus, combustion (and rust, for that matter) was an additive process rather than one that involved using up phlogiston, as Lavoisier proved by careful measurement and weighing of material—and ashes so produced—consumed by fire. Lavoisier made many other contributions to chemistry, as well as to sciences such as meteorology and physiology, but he also got involved with a tax farming scheme. The French government sold the rights to collect taxes, which meant that tax collection was subcontracted out to people who collected more than even the heavy sums the government demanded, in order to turn a profit. Tax farmers were hated in pre-revolutionary France, and when the revolution became radicalized in the 1790s, the radicals quickly went after the tax farmers. In 1794 Lavoisier died on the guillotine, as did so many others in France.

Chemistry wasn't the only area of scientific study to be revolutionized during the eighteenth and nineteenth century by application of techniques derived from Newtonian physics. There was a great passion for the study of electricity and magnetism, involving those who are household names, such as Benjamin Franklin (1706–1790), and those who should be household names, such as Nicolas Tesla (1856–1943). However, more relevant to our current study is the way in which the study of living things began to take on the outlines of a genuine science—or sciences, one should say, as such biological subfields as zoology and botany began to emerge. This would have important consequences for both the developing of scientific medicine as well as the relationship between religion and science—not just biological science, but science more broadly—from the nineteenth century through today.

During the eighteenth century, those known as naturalists—meaning that they studied the "natural history" of the world—primarily focused on systematizing, naming, and categorizing plants and animals. As had been true of so many people of the past, many of those interested in understanding the natural world were religiously motivated. For example, the most important of these scholars was Carl von Linné, known most commonly by the Latinized version of his name, Linnaeus (1707–1778). Originally educated to be a Lutheran minister, he viewed the study of the natural world—with a focus on what we would now call botany—to be a religious calling. For him, he believed in the divine ordering of the universe, including things such as the economic arrangements of societies. Therefore, he felt his study of plants was in line with what he saw as the divinely ordained agricultural system of the Swedes. Furthermore, like other naturalists of his day, he was convinced that God had created all creatures as part of the Great Chain of Being, in which living things were arranged with the highest order (God, angels, then humans) at the top, all the way down to the lowest order of life, with no gaps in the chain where a being might exist in theory, but didn't in actuality. All beings that could exist did and always had, representing what scholars referred to as the fixity of species. Species neither appeared anew, nor did they disappear.

Anyone with even a passing familiarity with biology and the history of biological organisms is aware of the problems with the concept of the fixity of the species. After all, there are no longer dinosaurs or dodos roaming the Earth, but Linnaeus was unaware of any such problem—at least at first. His most important work, *System of Nature* (first published in 1735), which presented the first taxonomic system of binomial nomenclature, in which organisms are organized according to genus and species, each of which are given a Latin name, included the phrase "no new species" within it. At first Linnaeus applied this system only to plants, but in the 1758 version extended it to animals as well. But the 1766 edition included the most important, though also the subtlest change—it dropped the phrase "no new species." Subtle though that change may have been, it didn't go unnoticed, as theologians soon began to accuse Linnaeus of atheism. His son hotly defended his now deceased father against such charges, avowing that his father firmly believed the natural order to be the handiwork of God—but he also stated that God's ordering allowed for the development of new species over time. In this he may have been influenced by his contemporary, the French naturalist Georges-Louis Leclerc, the

Comte de Buffon (1707–1788), although he didn't say as much. Buffon wasn't shy about his almost entirely negative views of Linnaeus, which might be why the Swede never mentioned him, but Buffon's thirty-eight volume *Natural History* was one of the most widely read works in Europe among the intellectual set. In it, he proposed that life may have been generated spontaneously in the warm oceans of an Earth that was 70,000 years old, rather than the less than 7,000 years old that most agreed to be the age of the Earth, and that this life mutated and degenerated according to geographical forces it experienced.

It's difficult to overemphasize the importance of Buffon's ideas about the age of the Earth and that living creatures change substantially through external influences. Many rejected these ideas outright as atheistic, at least at first, but others saw them as self-evident and providing great explanatory power. One such man was Georges Cuvier (1769–1839), a brilliant Frenchman who was a devout Lutheran and instrumental in founding the Biblical Society in Paris in 1818—he even served as vice-president—but whose greatest importance was in laying the foundations for the fields of paleontology and comparative anatomy in works such as the *Essay on the Theory of the Earth* (1813) and *The Animal Kingdom* (1817). Among a great many other accomplishments he identified bones found in the United States as a form of extinct elephant-like animal he named mastodon and was one of the first to suggest that reptiles had once dominated the Earth. However, he also vigorously opposed theories of evolution (and I use the plural intentionally because in these early days there was more than one such theory), instead arguing that the changing forms of living creatures was the result of new acts of creation following catastrophes that destroyed entire species of plants and animals. It's not at all clear that his religious beliefs played a role in his opposition to theories of evolution, as he instead made only evidence-based arguments in favor of catastrophism (as his view of periodic destruction of species followed by new acts of divine creation is known) versus evolution. One can speculate as to the motivations of supporters of catastrophism, as Carl Sagan did when he wrote in *Broca's Brain* that this theory "began largely in the brains of those geologists who accepted a literal interpretation of the *Book of Genesis,*" but it's important to note that the concrete evidence available in the eighteenth and early nineteenth centuries in support of early versions of theories of evolution were no more compelling than the evidence in support of catastrophism.

However, as we'll see shortly, Cuvier's religious beliefs did sharply delineate exactly how far he was willing to allow his scientific speculation to go. That becomes clear when we look at the conflict he entered into with his fellow countryman, Jean-Baptiste Pierre Antoine de Monet, Chevalier de Lamarck, typically known simply as Lamarck (1744–1829). A man who lived a fascinating life, he was the eleventh child of an impoverished nobleman who rode the breadth of France to join French forces fighting the Prussians, and served with distinction in an action that wiped out most of his company when he was only seventeen. He went on to suffer a career-ending injury during his military service—the result of a childish joke on the part of a fellow soldier—causing him to return to university to study medicine. However, his true passion was botany, and at the urging of one of his brothers, Lamarck decided to end his medical studies in order to pursue that passion.

By the mid-1790s Lamarck had established his reputation as a botanist (while also negotiating the tricky politics of being of noble birth surrounded by revolutionaries who were killing thousands of the members of France's nobility) on the basis of careful observation and botanical classification, as published in his 3-volume *French Flora.* This work represented what his contemporaries (such as Cuvier) widely recognized as the appropriate approach for a naturalist of the day—collect as much empirical data as possible, and present analyses that stick closely to that data without engaging in speculation or imaginative system building. By employing this socially-acceptable approach, Lamarck first obtained a position as Royal Botanist, entry into the French Academy of Sciences, then a far more lucrative position as botanist at the Royal Gardens—which Lamarck wisely renamed the Garden of Plants in 1790 in order to dissociate it (and himself) from the monarchy during the French Revolution.

However, while working at the Garden of Plants, Lamarck began to study mollusks—the single largest phylum of marine animals—of the Paris Basin, a major geological region in France. Although he was originally convinced of the fixity of species, studying these creatures convinced him that species changed over time in some manner. In 1800 he presented a paper providing a broad outline of his ideas, and over the next three years he published three major works—a book a year—detailing his ideas about the transmutation of living organisms in great detail: *System of Invertebrate Animals* (1801—this work would be expanded to seven volumes between 1815 and 1822), in which he coined the term "invertebrate," *Hydrogeology* (1802), and most importantly, *Research on the Organization of Living Bodies* (1803). In these works he laid out his theory that animals develop from simpler to more complex forms over time, influenced by geography, environment, and the behavior of the animals.

In Lamarckian evolution, life developed spontaneously. In his 1809 *Zoological Philosophy* he wrote that "Nature, by means of heat, light, electricity, and moisture, forms direct or spontaneous generations . . . where the simplest of . . . bodies are found." These bodies are possessed of an innate "power of life," as he termed it, which causes them to become more complex over time. The more complex the creature—a complexity brought about by the "power of life"—the greater the effect on its biological structures, occurring through use or disuse. The most famous example he provided was that regarding giraffes—he argued that proto-giraffes had customarily stretched their necks in order to reach leaves in trees, producing minutely longer necks in both males and females over the course of the life of the animal. Their offspring would then have slightly longer necks than their parents, who would then stretch their necks during the course of their own lives to reach leaves, leading to minutely elongated necks that would be passed on to their offspring—this process was repeated over the course of generations until the development of the long-necked giraffes with which we are all familiar.

The immediate public reaction to Lamarck's work was one of resounding silence, but it wasn't because of the power of the

Catholic Church—which would be restored to prominence in France alongside the restoration of the monarchy in 1814—which "had made of its enemies its footstool" in the evocative phrasing of Andrew Dickson White. It was because Lamarck's approach was out of step with the basic principles of science of his day. Although Lamarck was a thoroughgoing deist, as were many French intellectuals, many saw his "power of life" as an unprovable, untestable, and unmeasurable supernatural force, for the quite reasonable reason that it was. Therefore, this "power of life" was beyond the realms of science, and since Lamarck stressed that it was more important to his system than the influence of use and disuse, many intellectuals more or less openly laughed at Lamarck and didn't engage with him from a scholarly perspective because they didn't view his work as worth the effort. As for the lesser mechanism of evolutionary change—use and disuse of organs—Lamarck offered imaginative arguments but little in the way of concrete evidence. Nor could he answer direct challenges, such as when it was pointed out to him that since the time Greeks and Romans first took notice of giraffes, these creatures showed no tendency to develop longer necks. If Lamarck was right about why their necks had developed as they had, why did they stop elongating?

By the time a scientist—Cuvier—finally decide to speak up openly against Lamarckianism, Lamarck himself had become old and blind and couldn't answer the challenge. And it was a challenge that was brought about in part because of religious sentiment—Cuvier rejected the idea that life emerged without God's guiding hand, and that nature lacked divine influence—and partly because of concerns about the social situation in France. Charles X (r. 1824–1830), an uncle to Louis XVI (r. 1774–1793) who had died on the guillotine during the French Revolution, was now a constitutional rather than absolute monarch, as pre-revolutionary kings of France had been, and he was not at all happy with that. From the time he succeeded his childless brother Louis XVIII (r. 1815–1824) to the throne, he constantly sought to reassert the rights and powers of the monarchy and the nobility against those of the Prime Minister, the parliament, and the middle class. In this atmosphere, Cuvier—like many French intellectuals—saw Lamarckianism as dangerously encouraging political and social radicalism. This was no idle fear, as the July Revolution would force Charles X from the throne in 1830.

Thus, when Cuvier attacked Lamarckianism, an attack that culminated in a two-month long debate between Cuvier and his former close friend and disciple of Lamarck, Geoffroy Saint-Hilaire (1772–1844), it was for a complicated set of reasons, rather than being as simple as a religious challenge to science. Furthermore, Cuvier was widely acknowledged to have won these debates, as he came prepared with better evidence and more logical arguments, which readily convinced the members of the French Academy of Sciences—who were already widely skeptical of Lamarckianism and the "power of life" that underpinned it. Thus, Cuvier had evidence on his side. Finally, any evaluation of this episode as an example of a collision between religion and science is complicated by the fact of the supernatural nature of Lamarck's "power of life"—as well as by the fact that as far as Lamarck was concerned, evolution was the fulfilment of a plan God imposed on nature at the time of creation, and thus he was motivated by a desire to uncover God's divine laws as expressed in nature. He was a deist, not an atheist.

In part, what the clash between Cuvier and Saint-Hilaire demonstrated was the conservatism found in French science. Scientists tend to be conservative by nature, as slow to surrender cherished beliefs as any other person is—Thomas Kuhn's *The Structure of Scientific Revolutions* is still the most readable and in some respects still the best account of the reluctance of scientists to give up one theory for a newer one—but the structure of the French Academy of Science tended to reinforce this conservatism, as members had to fit their work into the acceptable boundaries of what constituted proper science if they wished to become eligible for one of the pensions that senior members received. The English Royal Society was no less hierarchical, which sometimes acted to slow scientific development—one need only look to Newton's presidency of the Royal Society in order to find examples—but allowed for a broader range of members who didn't compete for pensions. Furthermore, the United Kingdom had experienced less social and economic disruption than France had, while also enjoying the greater economic growth that the Industrial Revolution and imperialism allowed for. Economic growth doesn't always lead to intellectual development—in the United States it seems to promote reality television more than it does an interest in science—but science can't occur in the absence of the money to pay for it.

Charles Darwin (1809–1882) is an example of the important relationship between economic support and science. His father, Robert Darwin (1766–1844) was in the first instance a successful physician in the small West Midlands town of Shrewsbury. However, he also invested in real estate as well as canal and roadbuilding schemes, earning himself a sizeable fortune in the process. On his mother's side of the family there was even greater wealth—his maternal grandfather was Josiah Wedgwood (1730–1795), potter to the Queen and founder of the enormously successful Josiah Wedgwood and Sons, still renowned for the quality of its porcelain (the company merged with Waterford Crystal in 1987 to become part of first Waterford Wedgewood, then WWRD Holdings Ltd. in 2009). What all of this means is that Charles Darwin grew up in a decidedly upper-middle-class household, and money would never be an issue for him. However, he also grew up in a household in which education (his father and paternal grandfather were both physicians) was highly prized, as was hard work and frugality.

Another thing Charles inherited, besides a comfortable level of wealth and a strong work ethic, was religious non-conformity. His paternal grandfather Erasmus Darwin (1731–1802) may have been an outright atheist, as well as an early proponent of an evolutionary theory similar to Lamarckianism in his 1794 medical work, *Zoonomia* and his 1803 poem *The Temple of Nature* (it's worth noting that Erasmus was also an abolitionist and a proponent of education for women). Charles' father, Robert, was at best indifferent to religion. He valued the role of the religion in social life, but he didn't see any personal need to attend church. What little contact Charles had with religion, then, was through his mother—he attended a Unitarian church with her as a child, and she engaged the Unitarian minister to tutor Charles when he was eight. However, in the following year his mother died and his father sent him and his brother to a boarding school, where he would remain until 1825.

Robert Darwin had decided that his sons would follow their father and grandfather into medicine, so in 1825 he sent both his sons to one of the best medical schools in the world—Edinburgh University. Charles was only sixteen, a bit young for college, but his father thought spending time with his elder brother (named Erasmus, after their grandfather) while attending the odd lecture here and there would help lay the foundation for his success in medical school. However, he soon met a bad influence—that of Georges Cuvier, through his *Discourse on the Upheavals of the Surface of the Globe*, which was published the year after Charles Darwin came to Edinburgh. This work presented Cuvier's ideas about speciation as related to his belief in catastrophism, and it provided the young Darwin with much food for thought and sparked a lifelong interest in natural history—the budding young field of biology, to be precise.

By 1828 it was clear to Darwin and his father that he wasn't cut out to be a physician. Perhaps aided by the fact that Erasmus did indeed pursue a medical degree (though he would decide that he preferred living off his inheritance rather than working as a doctor), Robert Darwin consented to allow Charles to leave Edinburgh for Cambridge so that he could prepare for a profession for which he was more suited—that of an Anglican clergyman. As Charles would later state, he had examined his heart and decided (at least at that time) that he accepted the dogmas of the Church of England.

At Cambridge Darwin worked toward his B.A.—there were no "majors" in those days, everyone worked toward the same B.A., which was effectively a degree in classics—but also became an avid collector of beetles, as well as an assistant to Professor John Henslow (1796–1861), who held a chair in both mineralogy and botany. Henslow was firmly opposed to theories of evolution, and convinced that study of the natural world could provide evidence of the hand of God at work in the world—and Darwin agreed with the older man, whom he idolized, on both counts. However, Henslow was also deeply interested in how variations in species occurred, and he needed energetic young men such as Darwin to aid in sample collection. Henslow would also introduce Darwin to another professor who would prove to be a tremendous influence on the young man, Adam Sedgwick (1785–1796). As a professor of geology, he thought that he could gather evidence in support of the idea that the Earth had developed its current structures as a result of catastrophes that had occurred in the past, which Sedgwick was certain would be evidenced in rock strata. In particular, he sought to refute the work of Charles Lyell (1797–1875), who argued that the Earth was incredibly old, and had developed its current structure through many small naturally occurring changes, which occurred in accord with natural laws over the course of an unimaginably long span of time. To Sedgwick, this was patent nonsense that flew in the face of scriptural assertions about Noah's flood, and it was in order to disprove theories such as Lyell's that Sedgwick traveled to Wales to gather evidence in 1831. In what would prove to be a momentous occurrence, Henslow convinced Sedgwick to take Charles Darwin, even though the young man would already have graduated by the time of the trip.

Darwin's father was tiring of his son's hobbies, but agreed to allow him to go to Wales for a fourteen night trip instead of a month-long journey to the Canary Islands that the young man had proposed. Robert felt it was time for his son to give up such foolishness and get down to the serious business of what my students now call "adulting," so he agreed to support his son's jaunt to Wales, as long as Charles would agree to begin his studies for ordination upon his return. There is every indication that Charles meant to follow through on this bargain, but during his time in Wales he only fell more in love with natural philosophy as he marveled at the sedimentary layers revealed in Wales and wondered at the fossils found in these layers. Therefore, his focus was already on his love of uncovering nature's secrets when he returned to Cambridge, finding a letter from Henslow inviting

Charles to travel on a two-year journey around the world aboard a ship captained by Robert FitzRoy (1805–1865). FitzRoy was an amateur naturalist himself, and he wanted a gentleman with similar interests to undertake the voyage with him. As such, FitzRoy had approached Henslow, who turned him down, saying that someone younger would be better suited to such a challenging task. Someone like Charles Darwin.

Robert Darwin was initially wholly against the idea. His son Charles would have to pay his own way—which meant Robert would have to pay—but that was hardly the problem. What was more of a sticking point was that Robert felt Charles should stop dithering and start adulting. However, this was clearly a once-in-a-lifetime opportunity, and Dr. Darwin came around and agreed to support his son one more time. Thus began what turned out to be a five rather than a two-year voyage, which would not only change Charles Darwin's life, but would in fact change the world. Over the next five years, Charles continued his reading, with the most important work he read being Charles Lyell's two-volume *Principles of Geology* (eventually expanded to three volumes by 1833). Although Sedgwick was dismissive of this work, it proved to be of enormous importance to the development of Darwin's thought, for it presented a view of the development of the Earth in which slow, gradual change brought about the Earth as we see it now. This theory came to be known as uniformitarianism as opposed to catastrophism. Uniformitarianism proposed that extinctions such as those Cuvier presented evidence for occurred slowly over vast spans of time due to changes that occurred gradually rather than in the sudden dramatic catastrophes catastrophists believed in.

What was important about Lyell's geological theory was that it suggested the Earth was far older than Darwin was accustomed to imagining, on the order of billions instead of thousands of years. Thus, when Darwin encountered varieties of animals that were quite near to one another, yet had been cut off from one another by geographic features—such as tortoises and birds on the Galapagos Islands—Lyell's theories indicated that enough time had passed that speciation could occur. Animals that had originally been of the same species, when cut off from one another on individual islands, had developed into entirely new species over the course of the millennia.

Darwin didn't develop this notion immediately, of course. Nor did he follow it to its natural conclusions right away. Giving up on the idea of ordination, upon his return he continued reading, with Thomas Malthus' (1766–1838) *An Essay on the Principle of Population* (first published in 1794, going through numerous revisions until publication of the sixth edition in 1830) being of particular influence on Darwin. Malthus argued that the population increased geometrically while the food supply only increased arithmetically, leading naturally to crises as the population outgrew its food supply. Though focusing on human populations, he did couch it in terms of humans as animals, and stated that nature created various checks—such as famine and disease—to avoid these crises of the food supply. To Darwin, this suggested that those least suited to their environment would die first, with those who were the best fit for their environment surviving long enough to pass on their traits. A changing environment would change the traits that were best suited to the new environment, leading to changes that would result in new species—given incredibly long spans of time, which Lyell's geological theories allowed for.

In addition to reading, Darwin also set about gathering data in a journey that shows us both how meticulous he was, as well as how much science has changed from his day to ours. These came from a variety of sources, including an eight-year study of both fossil and living barnacles that led to the publication of a two-volume study of these marine mammals as well as his own experiments with pigeon breeding (a popular pastime of gentlemen such as Darwin). During this period he published quite a lot, though he didn't publish about evolution. He also wrote up a 230-page set of notes on his developing theory of evolution, but he told few people about what he was really working on. It's important to note that this type of science is all but unimaginable in today's world, where research scientists are driven by university rules to publish or perish, or by corporate demands to produce results or face unemployment. The idea of spending twenty years devoted to a single research goal is something that is unthinkable in today's world. However, Darwin was financially comfortable enough thanks to the wealth available to men of his class in an increasingly prosperous nation, so he didn't have to worry about pursuing financial support or the responsibilities of a job other than writing and researching, as most modern scientists have to do. Nor did he have to worry about producing results for his employers, since his personal passion to understand the world drove his work.

In that twenty-year period, Darwin also faced numerous life changes and challenges. In 1839 he married his cousin, Emma Wedgwood (1808–1896), a charming, intelligent, and deeply religious woman who would give birth to ten children, three of whom would die young. Robert Darwin died in 1848, and though he left his son enough money to keep him comfortable for the rest of his life, the elder Darwin's death was a tremendous blow. That blow was nothing, however, compared to the death of Charles Darwin's eldest daughter, Annie. She died at the age of ten in 1851, never having fully recovered from a bout of scarlet fever. It was her death that killed Darwin's belief in a God that personally intervened in the lives of people, as he found himself unable to fathom how an omnipotent and loving God could allow the suffering and early death of his beloved child. However, Charles Darwin never became an atheist. He maintained that the material world and its natural laws were only explicable by the existence of a creator God. However, Darwin didn't see evidence of God interfering with those natural laws He had put in place, making Darwin a deist.

Eventually, Charles Darwin published his groundbreaking work, *On the Origin of Species by Means of Natural Selection*, in 1859. Although he did not specifically discuss human origins in this book, he did use evidence of homologous structures between humans and other animals—such as the way in which dog forelimbs and human arms have similar bone structures—to suggest a common ancestor. Furthermore, he stated that in reading *On the Origin of Species*, "Light will be thrown on the origin of man and his history." The popularity of this work surprised his publisher, and demand far outstripped the supply of 1250 copies in the initial print run.

Response to Darwin was mixed, which was not at all the result of any sort of simplistic religion versus science mentality, in spite of Andrew Dickson White's assertion that *On the Origin of Species* had

> come into the theological world like a plough into an ant-
> hill. Everywhere those thus rudely awakened from their old
> comfort and repose had swarmed forth angry and confused.
> Reviews, sermons, books light and heavy, came flying at the
> new thinker from all sides.

There were religious objections to Darwin's work, but the situation was far more complex than White's bombastic rhetoric would lead readers to believe. True, the bishop of Oxford, Samuel Wilberforce (1805–1873), did write a review of *The Origin of Species* in which he stated that "the principle of natural selection is absolutely incompatible with the word of God," but other theologians disagreed. The Anglican priest Charles Kingsley (1819–1875) wrote Darwin to say that it confirmed his personal belief that a fixity of species was insupportable, an idea he arrived at from a consideration of "the crossing of domesticated plants and animals." Kingsley publicly stated that Darwin's view of God as working through natural laws provided "just as noble a conception of Deity" as one in which God persistently intervened in His creation. Kingsley's views were far from uncommon among religious leaders. Writing in 1868, the one-time Anglican priest turned Catholic priest (and future cardinal) John Henry Newman (1801–1890), stated in regards to Darwin's work that

> It does not seem to me to follow that creation is denied
> because the Creator, millions of years ago, gave laws to
> matter. He first created matter and then he created laws
> for it — laws which should construct it into its present
> wonderful beauty, and accurate adjustment and harmony
> of parts gradually.

Other theologians were even stronger in their approval of Darwin. For example, Henry Ward Beecher (1813–1887) was a popular Congregationalist minister in New York, and he referenced Darwin's theory a number of times in sermons he gave in 1885 and published *Evolution and Religion* in the same year. An active social reformer who had been an abolitionist practically from the cradle—his sister was the author of *Uncle Tom's Cabin,* Harriet Beecher Stowe (1811–1896)—he believed that God had created humankind with the potential for perfectibility. As far as Beecher was concerned, God guided the natural selection that made evolution occur, thereby demonstrating His majesty and love of humankind. Granted, this type of teleological thinking—that a species becomes "better" and "more advanced" over time—went against Darwin's actual theory, but Stowe would have replied that God could manage evolution any way He wanted.

We should also be mindful that not all opposition to Darwin arose because of religious feeling. Some took a strictly scientific approach to the question of speciation. For example, the Scottish physicist William Thomson (1824–1907) was one of the leading figures in the development of the new science of thermodynamics, and he simply didn't think the Earth was old enough to allow for the length of time required for changes in the manner that Darwin posited. According to his calculations the Earth had only cooled to a habitable degree from its original formation as a molten body some 100 million years before present. We know now that his assumptions about the Earth as having resulted from a cosmic collision are wrong, and that he didn't know about internal radiation (radioactivity had yet to be discovered) that keeps the Earth from cooling in the way he assumed—but the point is that challenges to Darwin's theory such as Thomson's were the result of careful reasoning. We shouldn't automatically assume that anyone who didn't immediately grasp the importance and usefulness of Darwin's theory was irrational or driven by religious feeling. Nevertheless, most scientists in the United Kingdom quickly grasped that Darwin's theory of evolution by natural selection was something far different than Lamarckian evolution had been. It was a theory supported by both a tremendous amount of evidence as well as rigorous logical reasoning, with considerable predictive power.

I've previously emphasized that social and cultural factors play a much larger role in the development of science than is commonly understood, and the acceptance Darwinian evolution attained in the United Kingdom and the United States further demonstrates the importance of these factors. This was a period when the British Empire was reaching its peak, devouring huge sections of the globe as technological development seemed to be rising to ever-higher peaks. The middle classes and the elites were convinced of the power and worth of science, which was the cornerstone of British expansion. The British saw themselves as the pinnacle of world civilization, and views of history as leading to continual improvements of British society, culture, and economic power were common. Thus, this idea of evolutionary progress, which posited organisms developing increasing complexity (or at least that was the common understanding of Darwinian evolution) fit well with cultural beliefs of the day. Therefore, while Darwin's long-awaited study of human evolution, *The Descent of Man, and Selection in Relation to Sex*, published in 1871, shook up many people—including such champions of Darwinian evolution as the Harvard professor of botany, Asa Gray (1810–1888), who saw Darwin as removing any possibility of divine direction in this latest work—the core tenets of Darwinian evolution rapidly became central to all of the biological disciplines.

There was, however, a dark side to the way those such as Beecher read and understood evolutionary theory. Many in the West already believed that Western civilization was superior to all other forms, and some seized on Darwin's theory to justify this notion. Most significant among these was Herbert Spencer (1820–1903), who had actually endorsed evolution—though in its Lamarckian form—, seven years before the publication of *Origin of Species*. Once he read Darwin, though, he was hooked. He was convinced that this was the theory he'd been waiting for, to explain why some societies came to dominate the world, and why some people—and some classes of people—did better economically and in the realm of political power than others. The Darwinian model of competition—which Spencer termed a competition for the "survival of the fittest," a phrase Darwin later came to borrow—suggested to Spencer that nations such as the United Kingdom that could conquer other peoples, such as those of India and much of Africa, were entirely justified in doing so. In a brilliant example of circular logic, the fact that the British could conquer the Indians meant that the United Kingdom was more advanced and thus should conquer India, in order to bring the so-called benefits of civilization to the Indians. Never mind the fact that India had an advanced society when Europeans were trying to figure out how to make fire by banging rocks together—this form of Social Darwinism allowed the people of the West to feel all warm and fuzzy about conquering people of other civilizations. Furthermore, Spencer's Social Darwinism supported the view of economics that British—and American—economic elites most wanted to believe in—the idea that the poor were poor by virtue of their own failings and the only way to help them was to let nature take its course. Sure, that meant "the starvation of the idle, and . . . [the] shouldering aside of the weak by the strong," but that was nothing more than "the decrees of a large, far-seeing benevolence," as Spencer wrote in 1851. The poor had to be allowed to starve for their own good, according to Spencer.

Spencer was wildly popular in his day, though that popularity declined sharply after 1900, and a wide array of European and American elites accepted the Social Darwinism that he promoted. However, it must be stressed that Social Darwinism is in no way based on an accurate reading of Darwin. Charles Darwin had much to say about how living things developed over time, and that included humans, but he had nothing to say about the development of human societies or people within societies. However, it's not a diversion from our topic of the relationship between science and religion to take a moment and consider Social Darwinism, because one of the most important challenges to scientific approaches to knowledge construction would emerge in the late nineteenth century in response (in very large part) to what many saw as not only Darwinian evolution's inherent atheism, but also what some saw as the connection of this theory to immorality.

In order to understand that connection, we need to focus for a bit on the United States as the nineteenth century gave way to the twentieth. Religion has always had a special importance to the social fabric of the United States. This isn't because of the myth of the "pilgrims" (a term not used until the nineteenth century) who supposedly landed at Plymouth Rock (another part of the myth not to emerge until the nineteenth century). Rather, it was because so many of the economic and political elites of this young nation had come to America as religious refugees or in an effort to impose their own views of religion on what they viewed as a blank slate for their own social experiments. Many imagine people came to America seeking religious freedom, but we must remember that the freedom they were seeking was freedom for their own religion—not for the religion of others. Out of the original thirteen colonies, eight had an official religion, and in those colonies, dissenters—those who held differing beliefs from those of the official religion—were persecuted and in many instances faced legal penalties. We can see this in incidents such as when the leaders of the Massachusetts Bay Colony's banished Anne Hutchinson (1591–1643) for maintaining that worldly success was no guarantee that one was bound for heaven, among other points, or when political and social elites banned or strongly discouraged the celebration of Christmas (seen as a pagan holiday) in Puritan-dominated areas such as Boston from the 1600s through to the 1800s. That religious impulse had never waned in America, but in the late nineteenth and early twentieth century, the way in which it was expressed took on particular overtones for those who were alarmed by changes the nation was experiencing—particularly the changes occurring in the American South.

By the beginning of the twentieth century, changes in American culture were raising alarm bells for many traditionalists. A growing number of young people were moving to the cities, where they pursued careers very different from those their parents had practiced in agricultural areas. The anonymity of these cities allowed for a loosening of traditional gender and sexual customs, as it became easier to for young people to date without concern about disapproval from hovering parents or watchful neighbors. Many women began to dress in ways that allowed for more freedom of motion—while also showing more skin—and began to take up pursuits such as drinking alcohol and smoking cigarettes in public that had traditionally been seen as only appropriate for men, causing many to decry what they saw as the negative effects of being separated from the land. Popular music began to show the influence of African American culture, as Jazz increased in popularity, and art began to reflect cultural values that many traditional-minded people found shocking. In the American South, state legislatures eager to improve the economic conditions of their states increased funding for education and began to enforce compulsory attendance laws, causing attendance to swell and many Southerners to begin to fear that impersonal modern-secular forces were replacing the traditional private-family orientation of Southern culture.

Across the country, traditionalists were deeply concerned about these changes. Parents worried their children were going to be corrupted (are parents ever not worried about that?), many rural Americans looked at the lifestyles and habits of those who lived in cities with distaste—and shuddered at the thought of what would become of young people who moved to the cities—and there was a common, generalized concern about the breakdown of traditional values. Although there were many sources of such concern, one that was particularly prominent in the South was a fear sparked by exploding school attendance, that these institutions of public education would teach children values and ideas inconsistent with those held by their parents. Opinion pieces and letters in newspapers decried the rise of atheism that Southerners were sure was occurring, which was assumed would undermine every value these people held dear. It was in this atmosphere that the Bible Institute of Los Angeles (now Biola University) commissioned Amzi Clarence Dixon (1854–1925) and Reuben Archer Torrey (1856–1928) to edit what would become a twelve-volume set of works titled *The Fundamentals: A Testimony to the Truth*, typically known as *The Fundamentals*.

The Fundamentals contained ninety essays written by sixty-four different authors drawn from a wide range of Protestant traditions. Published between 1910 and 1915, the Bible Institute of Los Angeles distributed these works free of charge to teachers, professors, ministers, YMCA and YWCA secretaries, and a host of other people deemed to be of particular influence in American religious life. More than three million copies were distributed, and *The Fundamentals* have gone through multiple printings (it's still in print today, in a four-volume version, or versions that squeeze together all twelve volumes into a single massive book with very tiny print) and has proved to have a tremendous impact on American society and religious thought. The primary targets of authors of the essays in *The Fundamentals* are various elements of modernism such as higher criticism (a form of biblical analysis that focuses on the reading of the bible as a historical document), socialism, modern forms of Christianity such as Christian Science and Mormonism, and the perennial bogeyman of Protestants, Catholicism. The fiery Scotsman T.W. Medhurst (1834–1917) even wrote an essay titled "Is Romanism [an old-fashioned and intentionally pejorative term for Catholicism] Christianity?" This was a period when millions of immigrants came to America from all over Europe, with Irish immigrants being some of the most visible. Many in America expressed intensely anti-Irish sentiment, that included attacks on the predominant religion among these immigrants—Catholicism. It was anti-Catholic sentiment that fueled the reemergence of the previously defunct Ku Klux Klan in 1910.

At the heart of the concerns expressed by the writers who contributed to *The Fundamentals* was a concern that modern culture was leading people away from the truth, toward atheism and immorality—which would lead to the breakdown and collapse of society. In fact, many writers expressed it not as some future concern, but as a very present one—they saw changing gender, sexual, and cultural mores as not simply evidence of an evolving nation, but as proof that society was in the process of collapse. The editors and contributors to *The Fundamentals* were adamant in their insistence that the bible was inerrant and should be read literally. They believed that it was the divine word of God, and that within its pages could be found evidence of the approaching end of times promised in the biblical book "Revelation to John," often just known as Revelations. These writers firmly believed that we were living in the end times, and the changes visible in modern culture was evidence of the impending second coming of Christ.

Within the viewpoint of contributors to *The Fundamentals*, the "Decadence of Darwinism," to borrow the title of Henry H. Beach's (1833–1918) contribution, was simply one symptom of the breakdown of modern culture presaging the end of days. However, it's important to point out that Darwinian evolution was not a prominent topic within the pages of *The Fundamentals*. Instead, most essays focused on a defense of individual elements of belief, such as the virgin birth of Christ or that only God can forgive sins (in contrast to the Catholic beliefs about papal or priestly absolution of sins and purgatory). Furthermore, we should be clear that what drove writers such as Beach was a concern that acceptance of Darwinian evolution—which he did attack as, in his view, contradictory in its approach—undermined what he saw as the true Christian beliefs that were the cornerstone of society. In his essay on Darwinism, he wrote: "Darwinists have been digging at the foundations of society and souls . . . Natural selection is a scheme for the survival of the passionate and the violent, the destruction of the weak and the defenseless." Thus, his critique was less about the content of evolutionary theory, and more about what he perceived its negative social implications to be.

Finally, though *The Fundamentals* supplied the name for modern Christian fundamentalists, who have made opposition to evolutionary theory one of the cornerstones of their beliefs, contributors to the work were far from united in their opposition to evolution. George F. Wright (1839–1921) contributed three essays, including one titled "The Passing of Evolution." However, he was far from an opponent of evolutionary theory. Although Wright earned a Doctor of Divinity degree from Brown University and a law degree from Drury University, he worked as a geologist for the United States Geological Survey for eight years before taking up a position as a professor of New Testament language and literature at Oberlin Theological Seminary. He later took a professorship at Oberlin in the "harmony of science and revelation." In his essay on "The Passing of Evolution," he wrote that "'Evolution . . . is coming to be recognized as but a new name for 'creation.'" A life-long outspoken defender of Darwin's theory—as well as geological conceptions of time—he argued that understanding evolution as the procedure by which God created the diversity of species on Earth could only magnify the awe humans rightly felt as they viewed God's divine power—and that a refusal to accept the evidence in favor of evolution was a rejection of the intellect with which God endowed humankind, and was thus an irreligious act of irrationality.

What we need to understand about early twentieth-century fundamentalism is that it was neither overly concerned with evolutionary theory nor united in condemnation of it. However, by the time the final volume of *The Fundamentals* was published, World War I (1914–1918) was tearing Europe apart. American religious conservatives were deeply concerned by this outbreak of war, and they rightly feared America would soon be drawn into the conflict. Fears that the end of days was approaching mounted as well as a growing belief among American religious conservatives that the war represented evidence of the decadence of modern culture. By 1918 the concern that this conflict heralded the coming apocalypse as foretold in Revelations had reached such a fever peak that massive crowds flocked to first Philadelphia, then New York, to attend conferences on biblical prophecy, where speakers pronounced that the "Lord's prophetic Word is at this moment finding remarkable fulfilment," demonstrating the "nearness of the close of this age, and of the coming of our Lord Jesus Christ," as the program of the Philadelphia conference proclaimed.

One of the most popular speakers at these conferences was the teacher-turned-Baptist minister, William Bell Riley (1861–1947). As Bell wrote in his book, *The Menace of Modernism*, he believed evolutionary theory to be both wrong as well as a threat to civilization. He stated that Darwinian evolution should be eliminated from educational curricula, because "evolution is not a science; it is a hypothesis only, a speculation." Such a statement indicates that Bell—like many people today—either don't understand what a scientific hypothesis is, or refuse to use the term in the way scientists do. A scientific hypothesis isn't just a speculation, it's an educated assessment of previous evidence and theories, providing speculation upon what might or might not be an explanation for unexplained elements of evidence and/or theories. For it to be a scientific hypothesis, it must be built on evidence-based reasoning, and it must be both testable and falsifiable. In other words, it must be possible to devise a test of the hypothesis that can lead to failure. Therefore, the hypothesis that environmental conditions cause species to evolve in certain ways can be tested by (for example) breeding generations of fruit flies within different laboratory-produced environments. However, stating that God guides evolution is a religious belief, because it's not possible to test such a statement, much less design a test that could fail—though such a belief isn't necessarily in conflict with scientific approaches to knowledge. Darwinian evolution isn't a hypothesis in the scientific sense—it's a theory, which has nothing to do with guesswork. A theory is the most comprehensive form of model within science, indicating that the idea (evolution, in this case) has been tested and retested, and is both supported by multiple lines of evidence and demonstrates the ability to both predict and describe a wide array of phenomenon.

It's not clear whether Bell understood he was misusing the term hypothesis or not, though it's important for us to understand what such terms mean if we are to understand science. What is clear is that for him, evolution represented a much more existential threat than even the most ardent critics of evolutionary theory who contributed to *The Fundamentals* believed. He argued that "propagation of the faith" was "the only antidote to that infidelity which is forcing its way beyond the very altars of our churches, and which has already slimed our schools with its deadly saliva." With this view in mind, he promoted a new conference to be held in 1919, to go beyond biblical prophecy. Thus was born the first World Conference on the Fundamentals of the Faith, held in Philadelphia in the summer of 1919. More than 6000 attendees came to this event that Bell proclaimed was "of more historical moment than the nailing up, at Wittenberg, of Martin Luther's 95 Theses."

Bell's rejection of evolution began to gain traction with many religious conservatives, thanks in no small part to the trauma of World War I. Feeling that Germany was wholly to blame, numerous American religious leaders asked why the Germans had acted this way. For many, the idea that something was rotten within German culture came from the German development of higher criticism—which necessitated a rejection of biblical literalism—and that thinking was only reinforced as the writing of Vernon L. Kellogg (1867–1937) became more widely known in American religious circles. Although an entomologist by training, he rejected Darwin's theory of natural selection. In 1915 and 1916 he took a leave of absence from his post at Stanford in order to work for the Hoover administration in Brussels, Belgium, where he met with members of the German high command. He was horrified by their Social Darwinist beliefs, and became convinced that these beliefs—which he wrongly traced directly to Darwin—were emblematic of the decadence of modern culture, and the root cause of World War I. He wrote about his ideas in a book called *Headquarters Nights*, published in 1917, and by the 1920s conservative religious leaders in America were using it as evidence that evolution was synonymous with violent aggression, cultural decadence, and atheism.

It was this odd blending of distaste for what many Americans—not just religious conservatives—came to see as the shortcomings of German culture combined with fears unleashed by modernization and World War I (which was also seen as a product of the decadence of modern culture, it should be remembered) that created such a ready audience for messages such as that promoted by Bell that many American religious conservatives began actively to attack evolutionary theory. This manifested itself most obviously in the American South where five state legislatures banned the teaching of evolution outright, leading to the famous "Scopes Monkey Trial" of 1925 in Tennessee, in which a substitute science teacher was convicted and sentenced to a symbolic fine for the teaching of evolution. However, in the 1930s opposition to evolutionary theory became muted to such an extent that it seemed to be completely forgotten among American conservative

Christians, and most religious leaders around the world were equally silent on evolutionary theory. That changed by mid-century, when a growing number of religious leaders indicated a growing acceptance of evolutionary theory, beginning with Pope Pius XII's (r. 1939–1958) 1950 pronouncement that the Catholic Church didn't forbid research on evolutionary theory. By the end of the century cautious acknowledgement of the value of evolutionary theory to science shifted to strong acceptance for many. Popes John Paul II (r. 1978–2005), Benedict XVI (2005–2013) and Francis (2013–present) have all affirmed that evolutionary theory is consistent with Catholic dogma, so long as it is recognized that the soul comes from God. And it isn't only Catholics who accept evolutionary theory. In his 1997 autobiography, *Personal Thoughts of a Public* Man, the Southern Baptist Billy Graham (1918–2018) wrote,

> I don't think that there's any conflict at all between science today and the Scriptures. I think that we have misinterpreted the Scriptures many times and we've tried to make the Scriptures say things they weren't meant to say, I think that we have made a mistake by thinking the Bible is a scientific book. The Bible is not a book of science. The Bible is a book of Redemption, and of course I accept the Creation story. I believe that God did create the universe. I believe that God created man, and whether it came by an evolutionary process and at a certain point He took this person or being and made him a living soul or not, does not change the fact that God did create man. . . . whichever way God did it makes no difference as to what man is and man's relationship to God.

It really doesn't get much clearer than that.

What Graham is expressing support for is known as Theistic Evolution, which is the dominant position among most religious groups in the West, and a good many around the world. It's taught in Catholic seminaries, most mainstream Protestant seminaries, and is accepted by most major Jewish, Muslim, Buddhist, and

Hindu groups in the United States and Europe. That fact isn't always readily recognized, however, as those religious conservatives who reject evolutionary theory are often very vocal. On the other side of the equation, there are biologists—such as Richard Dawkins—who roundly reject Theistic Evolution, claiming that it takes the "natural" out of natural selection. However, that doesn't appear to be the dominant position; a 2015 Rice University study found that in the U.S. and Europe, roughly half of scientists believe in God or a higher power, with a minority of scientists (29% in the U.S. and 34% in the U.K, for example) seeing a conflict between religion and science. Though the researchers didn't specifically ask about evolutionary theory, since the vast majority of scientists (not just biologists, but scientists more broadly) accept evolutionary theory, that would be a flashpoint for any scientist who sees religion and science to be in conflict. Furthermore, some scientists have been extremely strong in their support of Theistic Evolution. Theodosius Dobzhansky (1900–1975) was a geneticist and evolutionary biologist who was one of the chief architects of the modern evolutionary synthesis, blending Darwinian natural selection and the genetics work of Gregor Mendel (1822–1884— he was, it should be noted, a Catholic friar) with modern statistical methodology, which provides the foundation for all modern biology and medicine. Dobzhansky was also a devout Eastern Orthodox Christian, and in his 1973 essay *Nothing in Biology Makes Sense Except in the Light of Evolution* he wrote,

> I am a creationist and an evolutionist. Evolution is God's, or Nature's, method of creation. Creation is not an event that happened in 4004 BC; it is a process that began some 10 billion years ago and is still under way . . . Does the evolutionary doctrine clash with religious faith? It does not. It is a blunder to mistake the Holy Scriptures for elementary textbooks of astronomy, geology, biology, and anthropology. Only if symbols are construed to mean what they are not intended to mean can there arise imaginary, insoluble conflicts . . . the blunder leads to blasphemy: the Creator is accused of systematic deceitfulness

Most biologists seem to hold a similar position regarding Theistic Evolution.

As with all scientific advancements, Charles Darwin developed his theory of evolution within specific social and historical contexts. These contexts are also what led to the widespread acceptance of his theory, an acceptance that included many prominent religious leaders of his day. However, there were those from the beginning who saw evolutionary theory as a challenge to important elements of their faith. But from the beginning the reception of such criticism was predicated upon broader cultural factors than a simple question of what religion one might be. Religious leaders, from pastors and priests to cardinals and popes, were more likely than not to take no stance on or even support evolutionary theory during Darwin's lifetime and in the period soon after his death. By the twentieth century there did develop a concerted groundswell of opposition among certain Protestants, but social and cultural concerns drove this opposition more than any deep engagement with evolutionary theory. Furthermore, even as this opposition was developing, religious leaders were coming out in defense of evolutionary theory, and today religious acceptance of this cornerstone of biological science is mainstream, whereas rejection of evolutionary theory is restricted to a vocal (and sometimes politically well-connected) minority.

Further Reading

Bannister, Robert. *Social Darwinism: Science and Myth in Anglo-American Social Thought.* Philadelphia: Temple University Press, 1979.

Burkhardt, Richard Wellington. *The Spirit of System: Lamarck and Evolutionary Biology.* Boston: Harvard University Press, 1995.

Clayton, Philip and Zachary Simpson, eds. *The Oxford Handbook of Religion and Science.* Oxford: Oxford University Press, 2006.

Desmond, Adrian J. and James Moor. *Darwin: The Life of a Tormented Evolutionist.* New York: W.W. Norton and Company, 1991.

Hart, Daryl G. *That Old-time Religion in Modern America: Evangelical Protestantism in the Twentieth Century.* Chicago: Ivan R. Dee, 2002.

Hess, Peter M.J and Paul L. Allen. *Catholicism and Science.* Westport: Greenwood Press, 2008.

Jones, Steve. *Darwin's Island: The Galapagos in the Garden of England.* New York: Little, Brown, 2009.

Marsden, George M. *Fundamentalism and American Culture.* Oxford: Oxford University Press, 2006, 2nd edition.

Melling, Phillip. *Fundamentalism in America.* London: Routledge, 1999.

CHAPTER 7

Science and Religion in Conflict?

As we saw in Chapter 6, church leaders and religiously committed scientists were as likely to praise Darwin's *On the Origin of Species* as they were to attack it, and many who challenged Darwin did so on methodological and scientific grounds rather than religious commitment. Nevertheless, during the 1860s there was a growing feeling on the part of some scientists that science and religion were—always had been, and always would be—locked in a struggle to the death. Foremost among proponents of what is now known as the Conflict Thesis are two men we've met many times in previous chapters, the chemist John W. Draper and the historian (and president and co-founder of Cornell University) Andrew Dickson White. Draper tended to split his time between scientific and historical writings (he wrote about the Civil War and Benjamin Franklin, as well as works on chemistry and the history of science) but he is best remembered for his *History of the Conflict Between Religion and Science*, published in 1874, just three years after Darwin applied his theory of natural selection to humans in his *Descent of Man*. White wrote on French economic history before making a sharp turn toward focusing on the historical relationship between science and religion, as shown by his 1869 lecture "The Battle-Fields of Science" and more significantly, his two-volume *A History of the Warfare of Science with*

Theology in Christendom published in 1896. Of the two, Draper's book enjoyed greater initial success, but White's would prove to be more influential in the long run, in part because it was neither as overt nor as strident in its anti-Catholicism as Draper's work and in part because he wrote like the historian he was, including extensive references that made his work appear credible. However, forces beyond a simple unbiased analysis drove both White and Draper, and just as White was putting the finishing touches on his two-volume denial that science and religion could cohabitate, a man was being born whose life and work would provide a reminder that such a viewpoint is more polemical than it is rational.

To understand what that means, and why we shouldn't evaluate either Draper's *History of the Conflict Between Religion and Science or* White's *A History of the Warfare of Science with Theology in Christendom* as unbiased historical accounts—which is unfortunately common—we need to consider both of them in the context of the times when they were written. For that, we need to go back a few decades to the 1840s when a remarkable flood of Irish immigrants began to enter the country. While that might seem like a non-sequitur given our focus on the historical relationship between science and religion in the west, it's necessary if we are to understand these two historical studies. It was the background of that immigration which drove both Draper and White's views.

In the nineteenth century the British government controlled Ireland from London, and in most ways treated the island nation as it did any of its other colonies. Ireland was, in fact, the first colony the English established abroad. The Irish had few political rights and a class of Anglo-Irish gentry controlled the country. Many of these people had lived in Ireland for centuries—Queen Elizabeth I (r. 1558–1603) had tried to push the Irish out of Ireland altogether, and encouraged the immigration of Scottish Presbyterians and English Anglicans to Ireland in order to build up a ruling class—but they were nevertheless separated from the Irish population by language, religious belief, and social customs. While there was some interaction, for the most part the

Anglo-Irish gentry were physically and culturally separated from the native Irish. This situation dated back to the reign of King Henry II of England (r. 1154–1189), though for many centuries the numbers of the English were few in Ireland and the Irish nobility continued to control most of the island. However, under Henry VIII (r. 1509–1547) and Elizabeth the English had gotten far more serious about their desire to control Ireland. The Irish nobility were driven from the land and the Anglo-Irish gentry and nobility assumed full control over Ireland—though the Irish periodically caused them problems.

No group of people ever like to think of themselves as bad people, though, and the English were masters at generating rationales for why their control of Ireland and the Irish was actually in the best interest of the Irish. As I detail in *The Impact of the English Colonization of Ireland in the Sixteenth Century*, the English took every opportunity to portray the Irish as lazy drunkards who were incapable of governing—or civilizing—themselves. Never mind that the Irish had a long tradition of poetry, music, and scholarship of their own—this was mainly in Irish Gaelic and thus wasn't seen as properly demonstrative of intellectual ability, because the Irish didn't master the forms of scholarship in Latin the English (and the rest of Europe) saw as truly intellectual. By the nineteenth century the English even argued that the Irish tradition of poetry was proof of their inability to govern themselves. In the English way of seeing things, the Irish penchant for poetry and storytelling was evidence that they were wooly-headed dreamers instead of rationalists capable of running their own country.

The English also took control of most of the land (and all of the best of the land) away from the Irish, converting much of it into raising the sheep that provided the wool that was so valuable on the European market—and after the eighteenth century, this wool acted as the raw material that kept English textile factories in business. That and the cotton the English bought from the Americans or forced the Indians (another subjugated people) to grow in place of the food they actually needed. Therefore, by the nineteenth century the Irish had precious little land on which to grow food for themselves. Fortunately, they had learned that the potatoes Sir Walter Raleigh (1554–1618) had introduced to the island could be grown on very little land, and when all parts of the potato were eaten it provided a rather nutritious diet. Thus, by the nineteenth century what little land the Irish had left to them was planted almost entirely with potatoes, which provided the vast majority of their food supplies.

The drawbacks of this dependence on monoculture became abundantly clear when a fungus attacked the potatoes the Irish depended upon in 1845. For the next seven years the potato crops in Ireland failed almost in their entirety. The British government expressed concern that providing relief for the Irish would breed a culture of dependency, and for the most part did nothing more than take note of how many Irish people starved. Private charities did try to help, but they were swamped by the massive need they faced and between 1 and 1.5 million Irish starved while another 1 to 1.5 million emigrated in order to avoid starvation. The United States had already seen an influx of hundreds of thousands of Irish in the 1820s and the 1830s, drawn by the promise of work on the canals, railways, and other construction projects, but in the 1840s the floodgates of Irish immigration to America were thrown wide open. Over 780,000 Irish immigrated into the United States from 1841–1850, and even when the effects of the potato blight subsided the crushing poverty and lack of land in Ireland continued to drive the Irish flight out of their country. Over 914,000 Irish immigrated to the U.S. between 1851 and 1860, and until 1910 some 400,000 Irish a year (with a bump of over 655,000 annually between 1881 and 1890). Another 350,000 Irish immigrated to the U.S. in the first two decades of the twentieth century, until the Emergency Immigration Acts of 1921 and 1924 threw the brakes on immigration and imposed a quota system, drastically curtailing the number of Irish immigrants allowed into the country.

Simmering anti-Irish sentiment in America—inspired by and borrowing from similar sentiments in England—came to a boil in the nineteenth century. In some ways negative views expressed about the Irish might sound odd to modern readers. For example, the Irish were thought of as a race, with animalistic, racialized characteristics that set them apart from Anglo-Saxons. English and American newspapers also portrayed the Irish as violent alcoholics, a "wild, reckless, indolent, uncertain and superstitious race," as the future British Prime Minister Benjamin Disraeli (1804–1881, P.M. from 1874–1880) put it in 1836. But otherwise, the ways that Americans opposed to Irish immigrants described them should sound very familiar to anyone knowledgeable about the immigration debate in the twenty-first century. U.S. Groups such as the Know-Nothing political party, which was aimed at keeping the Irish out of political office, castigated the Irish for not speaking English (Irish Gaelic was the primary language for a great many Irish, and the only language thousands of them spoke when they arrived in the U.S.), for having an alien culture incompatible with that of Americans, for concentrating themselves in Irish enclaves, and for taking jobs away from people born in the U.S. A primary element of the distaste that many felt for Irish culture was that the majority of the Irish coming to America were Catholic, or "papists" as they tended to be called in the nineteenth century. Throughout the nineteenth century many businesses openly stated that "Irish need not apply" to open positions, and in the 1860s and 1870s anti-Irish and anti-Catholic feelings were at a fever pitch, resulting in frequent anti-Irish cartoons and newspaper editorials as well as the occasional anti-Irish or anti-Catholic riot.

This was the atmosphere in which Andrew Dickson White was educated and grew to maturity. Born in 1832 in Homer, New York, White was undoubtedly fully aware of the rising number of Irish Catholics in New York City, which was where a high percentage of the hundreds of thousands of Irish coming into the country arrived—and where a good many stayed (by 1910 there were more people of Irish descent in New York City than there were people in Dublin). His father, Horace, had started out as a farmer but went on to become a successful and wealthy businessman and banker by the time Andrew was born. At the insistence of his father, Andrew enrolled at the small Episcopalian school, Geneva College, in 1849. The younger White found this school not at all to his liking, and withdrew in 1850, much to his father's displeasure.

Fortunately the two reconciled and Andrew White enrolled in Yale, where the philosophy professor (and Congregationalist minister) Noah Thomas Porter (1811–1892) became something of a mentor to the young man. White found great success as a student at Yale, being inducted into the secretive Skull and Bones society (a group that has included many powerful people, including Presidents Howard Taft, George H.W. Bush and George W. Bush) and editing *The Lit* (since renamed the *Yale Literary Magazine*). He even won the De Forest prize for public oratory in his senior year—which was, at 100 dollars, the largest prize of its kind offered at any educational institution—before going on to study at the Sorbonne and the University of Berlin after graduation. Returning to America, White earned his M.A. in history from Yale in 1856.

Upon completion of his studies (it was common for academics not to go on for a Ph.D. in the nineteenth century), White was a professor of English and history at the University of Michigan from 1858 until 1863. In that year, he left the university and was elected to the New York State Senate, where he met his fellow Senator, Ezra Cornell (1807–1874). Cornell was a self-taught and self-made man, a former Quaker who started out as a farmer before making his fortune as the founder of Western Union. Together the two men formed the idea of founding a new university in New York, one that would unusually be unaffiliated with any religious institution. Cornell had become disenchanted about religious orthodoxy, having been cast out of the Quakers for marrying outside the faith, and White had found the atmosphere at Geneva College—affiliated with the Episcopalian Church—both stifling and chaotic. Thus, both men had reasons to keep theology separated from higher education, but when legislation was advanced in the New York legislature in 1865 for the founding of what would become Cornell University, fellow legislators as well as leaders of denominationally affiliated colleges and universities attacked the idea of an absence of religious affiliation as a cover for infidelity. That same charge was leveled against White's plan to intentionally hire faculty from a variety of religious denominations and include a non-denominational chapel on the Cornell campus.

Nevertheless, White got his university by the end of the year when the governor signed into law the legislation founding Cornell, and White was named as the first president. However, charges of infidelity continued to be flung at him and he justifiably felt himself to be under siege. Decades later he wrote: "I stood for a time on the defensive," but "finding that this only provoked new attacks I determined to take the offensive," which he did with a lecture he delivered in December of 1869 at Cooper Union in New York City titled, "The Battle-Fields of Science." In this lecture, he described "some of the hardest-fought battle-fields" of what he termed "the great war" between religion and science. His chosen examples focused on many of those discussed in the pages of this book, including Giordano Bruno who was "burned alive as a monster of impiety" and Galileo who was "tortured and humiliated as the worst of unbelievers." His lecture ended with the latest martyrs in this war between science and religion, Cornell University and its president, Andrew Dickson White. Horace Greeley (1811–1872) published White's essay the next day in the *New York Tribune*.

Needless to say, this lecture and essay sparked an outcry among the religiously inclined, though it also inspired plenty of supporters to speak up on White's behalf. It also set White on an academic path that would absorb him for the rest of his careers, as he continued to carry out research in order to support his Conflict Thesis of a continuous warfare between science and religion. He published a short treatment of the topic in 1876 as *The Warfare of Science* and continuously published articles in magazines such as *The Popular Science Monthly* before finally publishing his magnum opus, a 900 page, two-volume study titled *A History of the Warfare of Science with Theology in Christendom* in 1896. By this point his position had evolved to the point that White focused his attack on what he termed Dogmatic Theology, which he saw as based on ancient myths and thereby holding back both science and a pure revealed religion. White, unlike Draper whom we'll consider in a moment, pointed out that when Protestantism arose leaders such as Luther and Calvin could be just as dogmatically opposed to science as any Catholic. However, the vast majority of *A History of the Warfare of Science* is directed against Catholicism, using examples such as the supposed opposition of the Catholic Church to dissection and of course attacks upon the primary figures of the Scientific Revolution. We've already dealt with the ways in which he was mistaken about Copernicus, Kepler, and especially Galileo, but it's worth noting just how strong White's vitriol was when discussing the Galileo affair. He wrote:

> The whole struggle to crush Galileo and to save him would be amusing were it not so fraught with evil. There were intrigues and counter-intrigues, plots and counter-plots, lying and spying; and in the thickest of this seething, squabbling, screaming mass of priests, bishops, archbishops, and cardinals, appear two popes, Paul V and Urban VIII. It is most suggestive to see in this crisis of the Church . . . on the eve of the greatest errors in Church policy the world has known, in all the intrigues and deliberations of these consecrated leaders of the Church, no more evidence of the guidance or presence of the Holy Spirit than in a caucus of New York politicians at Tammany Hall.

For those who may have forgotten their nineteenth-century American history, Tammany Hall was infamous for its corruption.

One could forgive White for devoting so much focus to Catholicism. After all, for much of history Catholicism was synonymous with Christianity and established religion in the West. However, he specifically targeted "Christendom" in his title, a term used to denote the period in the Middle Ages when the unifying force of European civilization was Catholicism. As a historian, White was fully aware of that fact. Furthermore, there's no mistaking the fact that his most vitriolic statements are always directed at Catholicism. Thus, while he seems to have at least made an effort at being balanced in his attack on dogmatic theology and established religion, there is still a strong anti-Catholic line of polemic throughout his work. That's understandable given the historical context in which White was writing, but unfortunately it's all too common to encounter those—often, but not always, scientists—who read White's *A*

History of the Warfare of Science with Theology in Christendom in a thoroughly uncritical manner.

As I mentioned earlier, Neil deGrasse Tyson references and cites White in his *The Sky Is Not the Limit: Adventures of an Urban Astrophysicist* when he states that religion and science, "as they are currently practiced," are inherently in conflict. The astrophysicist also cites White a couple of times in his *Death by Black Hole*. Jerry Coyne (1949–present), a biology professor emeritus at the University of Chicago, has also written about "reading the famous 840-page anti-accommodationist book *A History of the Warfare of Science with Theology in Christendom*," no doubt looking for evidence to support his stance against religion. Coyne wrote an essay titled "Science and Religion Aren't Friends" for *USA Today* in which he stated

> religious claims retreat into the ever-shrinking gaps not yet filled by science. . . . Science and faith are fundamentally incompatible, and for precisely the same reason that irrationality and rationality are incompatible. They are different forms of inquiry, with only one, science, equipped to find real truth.

Hopefully by this point readers will see the problematic nature of such assertions. Back to the matter at hand, as a historian I would strongly caution against using a nineteenth-century book on history as a supporting source. History, like all other fields of inquiry, is ever changing and much of what nineteenth-century historians thought of as true has since been called into question or overturned. Andrew Dickson White's two-volume study is a fascinating primary source about nineteenth-century beliefs and attitudes, but it's practically useless as a source for understanding the historical relationship between science and religion. That's especially true given that readers such as Tyson and Coyne don't seem to have read White very closely at all, focusing on the bits that validate their own beliefs while overlooking statements he makes that might not fit so well with a commitment to the idea that science and religion are utterly incompatible. After all, White ends his Introduction by writing "that in the field left to them—their proper field—the clergy will more and more, as they cease to struggle against scientific methods and conclusions, do work even nobler and more beautiful than anything they have heretofore done," concluding that he holds the "conviction . . . that Science, though it has evidently conquered Dogmatic Theology . . . will go hand in hand with Religion." That's hardly an "anti-accommodationist" approach.

Turning to John W. Draper, he enjoyed greater initial success, perhaps because his book was a better fit for the times. Draper demonstrated a far stronger anti-Catholic sentiment, which fit well with a late nineteenth-century America where many people were upset by the hundreds of thousands of Irish (and mostly Catholic) immigrants entering the country. However, it's likely that Draper's unconcealed anti-Catholic bias is why his book has fallen out of favor. And it should be noted that Draper's anti-Catholicism can't be explained as simple anti-immigrant sentiment, much less as a reasoned analysis of the relationship between Catholicism and science. As we'll see, there may have been a complicating factor of a personal nature at play with the way Draper saw Catholics and Catholicism.

John William Draper was born in 1811 in the small town of St. Helens near Liverpool, England, a city that received a large number of Irish immigrants. By 1851 there were 43,000 Irish living in Liverpool, which as Frank Neal, author of *Sectarian Violence—The Liverpool Experience* points out, was greater than the population of most towns in Ireland. The number of Irish people living in Liverpool would grow to 83,000 (almost ¼ the population of Liverpool) before the end of the century, the vast majority of whom were Catholic and lived in conditions of grinding poverty. Nevertheless, tensions between the Irish and English (who were almost entirely Protestants) grew, as the English accused the Irish of stealing their jobs while attacking the Irish for "superstitious" rituals. These tensions culminated in a five-day riot begun when Protestants attacked a Catholic procession marking a religious holiday. Although Draper would be long gone by this point, having left the area to attend the University of London in 1829 before sailing for the United States in 1832—along with his three sisters, and his wife—it's worth remembering that Draper grew up as the son of a Methodist minister living close to a city where anti-Catholic tension was omnipresent and at times led to violence.

Upon arrival in the U.S., Draper first studied medicine at the University of Pennsylvania, receiving his M.D. in 1836, before going on to take a position as a professor of chemistry and natural philosophy at Hampden-Sydney College in Virginia. He remained there until he took a position as a professor of chemistry at New York University in 1839, where he helped found the New York University Medical School in 1841. He remained there as a professor of chemistry until 1850, before taking on the role of president of the medical school, which he held until 1870. Draper did his most important work in the fields of chemistry, making improvements in photographic plates that allowed him to introduce the use of photography into the field of science (and he is also credited as taking the first portrait of a person), and the study of radiant energy, discovering in 1842 that only those light rays that are absorbed can produce chemical change, which is now known as the Grotthuss-Draper law. Draper also became the first president of the American Chemical Society in 1876. However, for good or ill it is his historical work on religion and his views regarding Catholicism for which Draper is best remembered.

In 1860, seven months after the publication of Darwin's *On the Origin of Species,* John W. Draper presented a paper at the Oxford University Museum with the splendidly Victorian title, "On the Intellectual Development of Europe, considered with reference to the views of Mr. Darwin and others, that the progression of organisms is determined by law." However, his paper was quickly forgotten (as most conference papers are, to be perfectly honest) thanks to an exchange between the Anglican bishop Samuel Wilberforce (1805–1873) and the biologist Thomas Huxley (1825–1895). Wilberforce was a respected historian and member of the Royal Society, but also rather conservative and thus was opposed to Darwin's theory of evolution by means of natural selection. Huxley, on the other hand, was a biologist and anatomist who held the post of Professor of Natural History at the Royal School of

Mines (since 1907 part of the Imperial College of London) who had been initially hesitant about Darwinian evolution before coming to grab onto the theory for its explanatory power with such ferocity that he became known as "Darwin's Bulldog" for his defense of the theory. Huxley relished this title, even though he never expressed full support for Darwin's gradualism or the mechanism of natural selection.

Word had gotten out that Wilberforce intended to speak out against Darwin at the Oxford University Museum and hundreds of people had come to see the expected clash. After Draper's talk, which was universally regarded as overly long and boring to the point of being coma inducing, the moderator—Darwin's friend and mentor John Henslow, who it should be remembered was both an Anglican priest as well as a botanist and geologist—called upon several other attendees before it was Bishop Wilberforce's turn. When called upon, rather than making religious objections to Darwin's theory, Wilberforce instead took Darwin to task for relying too much on hypotheses rather than sticking closely to observable, concrete evidence. This approach was unlikely to have been a surprise, given that Wilberforce had written a 17,000 word review of Darwin's *Origin of Species*, which had recently been published in the *Quarterly Review*. This review is well written and demonstrates a thorough understanding of Darwin's work, while highlighting a central methodological debate in nineteenth-century science, whether scientific work should adhere strictly to rules of deduction or if induction was also allowable. Deductive reasoning flows from the general to the specific, in which scientists draw inferences from observable evidence. A rather simple example is the inference that neon is stable, drawn from the observations that noble gases are stable and neon is a noble gas. Deductive reasoning is meant to eliminate any possibility of doubt that a conclusion might be incorrect, and many nineteenth-century scientists argued that deductive arguments were the only allowable forms of argument in science. However, inductive arguments work the opposite way, starting with specific observations and leading to broad generalizations. That's exactly what Darwin was using when he stated, for example, that,

> For myself, I venture confidently to look back thousands on
> thousands of generations, and I see an animal striped like
> a zebra, but perhaps otherwise very differently constructed,
> the common parent of our domestic horse, whether or not
> it be descended from one or more wild stocks of the ass,
> heminus, quagga, or zebra.

Darwin had observational evidence at hand that animals like the wild ass and zebra are homologously very similar, and he inferred inductively that there must be a common ancestor, from which these modern forms had descended through the process of natural selection. For Wilberforce, as well as for many scientists, including no lesser a light than Adam Sedgewick, such inductive leaps were unwarranted. Therefore, while Darwin's ideas seemed worthy of interest suggesting possible lines of future research, for scientists and scholars who privileged deductive over inductive reasoning, what Darwin was doing wasn't science at all.

Huxley responded by emphasizing the explanatory power of Darwin's theory, which for him made it obviously the correct approach to understanding speciation, but most observers noted Huxley's high and reedy voice didn't carry very well, in opposition to Wilberforce's confident baritone that had earned him the reputation of being one of the best public speakers in Britain. Wilberforce may or may not have tossed out a humorous gibe at Huxley, asking whether he considered himself descended from a monkey on his grandfather or grandmother's side, and Huxley may or may not have responded that he'd rather be related to a monkey than to a man such as Wilberforce who attempted to use his oratorical powers to quash free discussion of a matter of truth. Attendees differ as to their memory of that detail. However, what is certain is that both sides saw themselves as the victors, the disagreement was over what sort of methodology is appropriate in the practice of science, and finally that this event was all but forgotten until it was resurrected as a tribute to Huxley following his death in 1895.

The point is, this debate was not the clash of "religion versus science" that it is sometimes portrayed as, and it was, in point of fact, a moment of very little importance. It wasn't really a debate at all, but an exchange that occurred after Draper presented his paper. In some ways it was likely to be of greater importance because of the influence it had on Draper than any impact it had in the history of science, for this conference in 1860 seems to have been the moment in which he became more interested in the history and philosophy of science than its practice.

In this new direction, John W. Draper didn't take an unbiased view of the relationship between science and religion. Instead, both public and private concerns colored his efforts to understand the history the two realms of inquiry shared. In the public sphere were the anti-Catholic feelings that existed in both the England of his youth and the United States that had become his adopted home, partially as a byproduct of anti-Irish immigrant sentiment and partially as the result of a long history of animosity toward Catholicism that was the legacy of English versions of Protestantism. In the private sphere, the issues were far more personal. In 1853 his sister Elizabeth had come to live with him for a time. She was a convert to Catholicism, and when Draper's eight-year-old son William was ill and lay close to death, she hid his favorite book—a Protestant devotional tract—and didn't admit to it until after William's death. John was furious and immediately kicked her out of his house. The incident left him with considerable animosity toward Catholics for the rest of his life.

This animosity revealed itself in Draper's first foray into history, his 1862 two-volume study of the *History of the Intellectual Development of Europe*. In this book, when he considers the Middle Ages, though he does have some good things to say about early popes—such as Pope Gregory I (the Great) (r. 590–604), whom Draper calls "virtuous," though he dispenses with discussion of Gregory in a few paragraphs—he spends far more time dwelling on the personal failings and scandals that were sometimes associated with the papacy. Draper spends pages discussing popes who had their opponents blinded or killed, men who were "so foul, so execrable" that their successors shuddered to mention their names. While almost nothing is said of popes such as Innocent III (r. 1198–1216), men who lived exemplary lives dedicated to service, Draper seems unable to say enough about periods when less capable—and upstanding—men held the papacy, whose actions were enough to make "the historian shut the annals of those times with disgust." These were times when "the heart of the Christian" might "sink within him at such a catalogue of hideous crimes." Draper draws a direct parallel between the bad popes of the tenth century and a period "several centuries later . . . when public opinion [did] come to the true and philosophical conclusion—the total rejection of the divine claims of the papacy." Never mind that "public opinion" never turned against "the divine claims of the papacy," or that only a biased view of events can assume that rejection of papal claims to power over the church can only be "true and philosophical" if viewed from outside the Catholic perspective—and the majority of Christians have been Catholics since sometime during the fifth or sixth century, a situation that remains true today. For Draper, it was self-evident that the papacy is an "evil institution" and that Catholicism is unworthy of respect.

Although Draper's *History of the Intellectual Development of Europe* was meant to be broad in its focus it was held together by Draper's belief that this was a "physiological argument respecting the mental progress of Europe." Draper confidently tells the readers that "[i]t is the special object of this book to demonstrate the proposition that social advancement is as completely under the control of natural law as is bodily growth." In other words, the progressive development of European intellectual traditions occurred in an upward trend, following rules of progress much like Draper's understanding of Darwinian evolution. And part of that progress was a rejection of what Draper saw as the errors of the Catholic Church in favor of views that are obviously "true and philosophical."

Draper's first attempt at history enjoyed a fair bit of success, in spite of its "reckless looseness of style, and utter absence of logical sequence," as one reader put it in his Oct. 25th, 1863 letter to the editor of *The New York Times* in reference to their review of *History of the Intellectual Development of Europe*. After writing a three-volume *History of the American Civil War*, which he published between 1867 and 1870, Edward Livingston Youmans (1821–1887) approached Draper with the idea that he write a volume on science and religion for his "International Scientific Series." Draper agreed, but had numerous professional obligations, so rather than starting from scratch he condensed portions of his earlier *History of the Intellectual Development of Europe,* wrote a new preface and three chapters focused on recent events and titled this book *History of the Conflict between Religion and Science* (1874). This book proved to be enormously popular, going through fifty printings in the U.S., twenty-one in the U.K, and was translated into nine languages— the Spanish version was placed on the papacy's *Index of Forbidden Books*—and whereas Draper's previous work contained notes of anti-Catholicism, this book can be called with fairness anti-Catholic through and through.

For Draper, it was a simple question of rationality versus superstition, and events occurring in the years leading up to his *History of the Conflict* drove him to see the Catholic Church—which he already viewed in a strongly negative light—as the promoter of superstition and a steadfast opponent of science.

And recent events had pushed him to hold an even more strongly negative view of the Catholic Church than that derived from his personal experiences and the anti-Catholic views common in the U.S. In 1864 Pope Pius IX (r. 1846–1878), worried about what he saw as the fast pace of changes in modern society, issued a document known as the *Syllabus of Errors.* The *Syllabus* attacked everything from rationalism and naturalism to bible societies and socialism. Men such as Draper saw this document as a strong attack on modernity—and in many ways it was—and science, which is far more debatable. The *Syllabus* does soundly reject the idea that "Human reason, without any reference whatsoever to God, is the sole arbiter of truth and falsehood, and of good and evil," (statement #3) but as Cardinal Newman would point out in 1874, the *Syllabus* is a collection of statements worthy of condemnation, in the eyes of Pope Pius IX, and those statements can't be fully understood without an examination of the documents from which they're drawn, in light of the context of a full understanding of theology. Thus, the *Syllabus* doesn't condemn reliance on human reason. Instead, it condemns the idea that human reason *alone* is the arbiter of truth. While many might disagree, none should be shocked or dismayed to hear the pope condemn the idea that human reason can act as the arbiter of truth without reference to God or his plan. Nevertheless, Protestants almost uniformity expressed deep anger about the *Syllabus*, and many scientists stated that it was a direct rejection of science. Therefore, Draper not only had personal reasons for harboring animosity toward Catholicism, at a time when most Americans were heavily prejudiced toward Catholicism, but he also had current events in mind when he sat down to write.

As if that weren't enough, in 1870 the First Vatican Council affirmed that the pope is infallible when making an official pronouncement (curiously, this position was deemed a heresy throughout the Middle Ages). In response, Draper wrote in his *History of the Conflict between Religion and Science* that: "For Science the criterion of truth is to be found in the revelations of Nature: for the Protestant, it is in the Scriptures; for the Catholic, in an infallible Pope." Throughout this book, Draper portrays the papacy as corrupt, the Catholic Church as repressive toward science, and Catholicism as riddled with superstition. The only times Draper has positive things to say about religion is when he refers to "Mohammadism," Islam, which he saw as a force for the promotion of science.

As I've noted many times previously, Draper's work contains a great many myths, outright falsehoods, and strongly held biases masquerading as reasoned evaluation of the evidence. For all of these reasons the book eventually fell out of favor, but it was also because of these reasons that it was perfectly suited for its time. Anti-Catholicism had been a persistent strain among the English since the time that Henry VIII had broken from the Catholic Church in the 1530s and Elizabeth I had ordered dozens of Catholic priests burned for their real or imagined efforts to subvert her rule. That anti-Catholicism had only been strengthened by the English attitudes toward the Irish, as a useful element of rhetoric for why Ireland should be under English control. When the Irish then began to immigrate into England in large numbers due to poverty and famine caused in large part by English misrule of their island, English anti-Catholicism reached a fever pitch, providing a workable template for Americans who were angered by the waves of Irish immigrants. Given this atmosphere, it's no wonder that Draper's intensely anti-Catholic *History of the Conflict between Religion and Science* proved popular. And yet, just twenty years after this book was printed, right around the time White published his *A History of the Warfare of Science with Theology in Christendom*, a man was born who would yet again prove that equating Catholicism (or Christianity) with opposition to science is illogical.

Georges Lemaitre (1894–1966) was born in Charleroi, Belgium. He initially received a typical-for-the-day classical education at the Jesuit run Collège du Sacré-Coeur, before entering the Catholic University of Leuven to study Civil Engineering at the age of seventeen. In 1914 he left off his studies to in order to take a commission in an artillery company—where he could put his training in mathematics to good use—and served throughout World War I. At the end of the war he received the Military Cross with Palms for courage in action. He then returned to Leuven to study math and physics. However, inspired by one of his teachers, Cardinal Desire-Joseph Mercier (1851–1926), he decided to devote his life to the priesthood. He earned a Ph.D. in mathematics in 1920 and Cardinal Mercier personally ordained Lemaitre a Catholic priest in 1923. Wishing to encourage Lemaitre's interests in mathematics and science, Cardinal Mercier then sent the younger man on to continue his studies at the University of Cambridge, which would prove truly momentous.

Cambridge—where it should be remembered Newton had once been on faculty—had many talented professors, but none would prove to be more important to Lemaitre's future than Arthur Eddington (1884–1944). Perhaps the deep faith the two men shared aided in drawing them together—Eddington was a devout Quaker who had almost gone to prison during World War I for his refusal to take up arms—but if so, it was their mutual interest in mathematics and an intense desire to understand the universe that acted as the cement for their relationship. Eddington had been on faculty at Cambridge since 1913 and in 1914 had been named chair of the Cambridge Observatory. Perhaps just as importantly, during WWI he became secretary of the Royal Astronomical Society, and it was in this capacity that he received letters from a Dutch physicist about Albert Einstein's (1877–1955) general theory of relativity, a theory Carl Sagan rightly calls "epochal" in episode 8 of his iteration of *Cosmos*. Einstein had been working on this theory between 1907 and 1915, and in 1916 published a paper describing the theory and providing an explanation for gravity, as a warpage of the space-time continuum. Einstein's theory overthrew the classical model of physics Newton had developed over two centuries previously and may well be one of the most important scientific theories of all time, but it is also one of the most complex. Few had the mathematical skill to understand it fully, but Eddington was one of those few. He quickly became a champion of Einstein's theory and set out to prove its accuracy, doing so with one of the clearest examples of what Isaac Newton had termed a "crucial experiment," a single experiment capable of proving or disproving the value of a theory. This occurred in 1919 when Eddington recorded the positions of stars around the Sun during a solar eclipse from the vantage point of Principe Island off the coast of Africa. By doing so, Eddington was able to prove definitively that an effect known as gravitational lensing, where gravitational forces cause light to bend, occurs precisely the way Einstein had predicted. Eddington would go on to make many contributions to astronomy, which would prove to be inspirational for Lemaitre. Just as importantly, though, Eddington was a strong supporter of the idea that scientific knowledge should be made as widely accessible as possible.

After spending 1923 studying with Eddington, Lemaitre traveled on to Harvard College Observatory and Massachusetts Institute of Technology, where he studied throughout 1924, and registered as a doctoral student at M.I.T. Upon completion of this whirlwind intellectual tour, Lemaitre returned to Belgium. There he took up a post at his alma mater, the Catholic University of Leuven, where he remained for the rest of his life. Thanks to his contact with Eddington, Lemaitre had become fascinated with Einstein's theory of general relativity, and upon his return to

Leuven, Lemaitre dove into work on this theory. This work led to his discovery of solutions to Einstein's equations that suggested the universe was expanding, rather than static. Einstein had actually realized this as a possibility of his work, but had introduced a fudge factor into his own equations because he thought the idea of an expanding universe was logically inconceivable. Lemaitre, however, didn't shy away from the idea and in 1927 he published what would prove to be a groundbreaking article with the unwieldly title (in English translation), "A homogeneous universe of constant mass and growing radius accounting for the radial velocity of extragalactic nebulae." Although the title is a mouthful and all but indecipherable for anyone who doesn't already know what Lemaitre is driving at, the paper makes a strong argument in favor of an expanding universe, while including what would later become known as Hubble's Law (that the speed of objects in an expanding universe is directly proportional to their distance from us) and the Hubble constant (a value for the rate of expansion).

As important as this paper was, it received almost no notice because Lemaitre published it in a rather obscure journal read by hardly anyone outside Belgium—though it did get the attention of one rather important reader, Albert Einstein. The two men met in 1927 when they attended what is one of the most impressive gatherings of all time, the 5th Solvay Conference in Brussels. These conferences, named after the Belgian industrialist Earnest Solvay (1838–1922), are invitation only and in 1927, of the twenty-nine attendees seventeen were either already Nobel Prize winners or would go on to receive this prestigious award, including the first woman to win a Nobel, Marie Curie (1867–1934), who won not one, but two Nobels, in two different sciences. When Einstein met Lemaitre at this conference, he stated that he was impressed with the younger man's work—but also stated that he couldn't accept his conclusions regarding an expanding universe. Einstein told Lemaitre, "Your calculations are correct, but your physics are abominable."

This type of rebuke must have been quite unsettling for a man so invested in research on relativity as Lemaitre was—as shown by the title of the dissertation he returned to the U.S. to defend at M.I.T. later in 1927, "The Gravitational Field in a Fluid Sphere of Uniform Invariant Density According to the Theory of Relativity." However, his defense went smoothly and Lemaitre returned to Leuven, now with a second Ph.D.—this time in physics— in hand, more determined than ever both to tease out the mysteries of the universe as well as gain recognition of his ideas. In 1929 Edwin Hubble (1889–1953) published the results of a decade of observations, including evidence that the speed of objects moving away from our point of observation is directly proportional to their distance from us—concrete evidence in support of Lemaitre's theory of an expanding universe. At this point Lemaitre sent his little-known paper to his former professor at Cambridge, David Eddington, who immediately recognized its value and had it translated (except, curiously enough, for two pages that detail what is now known as Hubble's Law and the Hubble constant) and republished in the *Monthly Notices of the Royal Astronomical Society*, in 1931.

Up to this point, it's not clear whether or not Lemaitre understood the full implications of his work. If *all* the matter in the universe appears to be receding away from us, unless one imagines that the Earth is the literal center of the universe, the only logical explanation is that all observable matter is part of an explosion that set everything rushing away from everything else. Furthermore, Lemaitre had the calculations to show that the universe is expanding. One need only rewind those calculations, as it were, working them backwards, in order to see that all matter had eventually occupied a single point in space. The first clue that Lemaitre had come fully to realize the implications of his theory is from a 1931 meeting of the British Association in London, which focused on the relationship between spirituality and the observable universe. At this meeting Lemaitre first noted that his theory suggested that the universe began with all matter packed into a single point, which he referred to as the "primeval atom," or "the Cosmic Egg, exploding at the moment of the creation." Lemaitre worked out the mathematics and published a paper on this idea later in the year, in the widely read journal *Nature*.

The immediate response wasn't good. However, opposition to Lemaitre's theory didn't come from religious quarters. In fact, the Catholic Church in particular rejoiced at the Belgian priest's findings, with Pope Pius XII (r. 1939–1958) proclaiming in 1951 that the theory is evidence of a "transcendental creator," in spite of Lemaitre's protestations that his theory had nothing to do with religion. Instead, resistance came from secular-minded scientists who saw the theory as too reminiscent of a religious idea. After all, if everything in existence came from the explosion of a "primeval atom," many saw that as an implication that this atom required a creator—just as Pius XII would later state. Eddington stated that the idea was "repugnant" and Einstein rejected the notion out of hand. The strongest opposition to Lemaitre's theory, though, came from the astronomer Sir Fred Hoyle (1915–2001). An outspoken atheist and controversialist, Hoyle saw Lemaitre's theory as not at all scientific, nothing more than a religious idea dressed up as science. It was, in fact, Hoyle who coined the term "the Big Bang theory" during an extremely derisive talk he gave on the British Broadcasting Corporation (BBC) in 1949. In another appearance on the BBC, Hoyle stated "The reason why scientists like the 'big bang' is because they are overshadowed by the Book of Genesis. It is deep within the psyche of most scientists to believe in the first page of Genesis."

Fortunately, not all scientists were as dogmatic as Fred Hoyle. In 1933 Lemaitre and Einstein traveled together to a conference at the California Institute of Technology in Pasadena, California. Wearing the standard attire of a Catholic priest, including the distinctive "dog collar," as he always did, Lemaitre presented his talk on what he called the primal atom or cosmic egg theory. This wasn't a standard conference presentation, but a high-level exposition of the mathematics of theoretical physics, in which he laid out both the argument of his theory and the evidence for it. When he finished, Einstein—who up to this point had rejected Lemaitre's theory as illogical and overly religious in its conception—stated, "This is the most beautiful and satisfactory explanation of creation to which I ever listened." A reporter for *The New York Times,* Duncan Aikman (1890–1955), noted the "profound respect and admiration" that Einstein and Lemaitre had for one another, and went on to write in what could be seen as a direct rebuke of the work of both Andrew Dickson White and John W. Draper that,

> 'There is no conflict between religion and science,' Lemaitre has been telling audiences over and over again in this country . . . His view is interesting and important not because he is a Catholic priest, not because he is one of the leading mathematical physicists of our time, but because he is both.

By the time Aikman made this highly telling observation, the scientific world was coming to accept Lemaitre's theory, and within a few years all but the most obstinate scientists—with Fred Hoyle being the most obstinate of all—would come to see what is now known as the Big Bang Theory as the most likely description of the formation of the universe. Further articulated by the Russian theoretical physicist George Gamow (1904–1968)—who spent most of his career at George Washington University, after defecting from the Soviet Union in 1931—scientists in a wide variety of fields have gathered an impressive array of evidence in support of the Big Bang Theory, making it difficult to dispute from a scientific standpoint.

Of course there are some who reject the Big Bang Theory on religious rather than scientific grounds. Beginning with George McCready Price (1870–1963), a Canadian Seventh Day Adventist, some biblical literalists insist in the face of all evidence that the Earth is only thousands of years old, rather than the billions the scientific evidence supports. Although this "young Earth creationism" started within the Seventh Day Adventist tradition, it has spread and influenced a wide variety of fundamentalist Christians, particularly in the United States where 51% of people express doubts about the validity of the Big Bang Theory, according to a 2017 AP-GfK Poll. Some of this doubt may come from the fact that many people simply have a poor understanding of the theory and find it difficult to wrap their heads around. Still, it's ironic that

Georges Lemaitre had to fight against atheists and agnostics who felt that he, as a priest, was presenting a religious idea dressed up as science, and now that more or less all scientists have come to accept the validity of the Big Bang Theory, opposition has arisen to his theory on the grounds of religion. It's interesting that neither Carl Sagan's 1980 version of *Cosmos* nor Neil deGrasse Tyson's 2014 reiteration of the show mentions Georges Lemaitre, given the emphasis on astronomy and cosmology on both series. Perhaps a priest who wore clerical attire while conducting science didn't fit well with the conception of either series.

However, any simplistic view that those who are religious reject the Big Bang Theory and those who are scientific in their approach to the world accept it is utterly flawed. In May of 2017 the Vatican hosted a conference at the Vatican Observatory, which Pope Leo XIII (1878–1903) founded in 1891, in order to celebrate the life of Georges Lemaitre. The Vatican Observatory was far from the first observatory the Vatican had established: that honor could be claimed by the Observatory of the Roman College, which had been founded in 1774, demonstrating the Catholic Church's longstanding interest in the promotion of astronomy. Scientists from around the world came to give papers on such topics as black holes and gravitational waves, in honor of the man who said that Christians believe in a God who created the universe, but "our science tells us how he did it." Such a gathering makes Draper's notion that Catholicism in particular is inimical to science look rather foolhardy, and given Lemaitre's example—and the opposition that Hoyle made to his theory—White's idea that opposition to science derives from Christianity (or dogmatic theology, as he put it) is simply insupportable. Certainly, some people of faith oppose scientific ideas, but there's no reason to imagine that such opposition springs naturally from their faith. In reality, it's far more likely to be the result of a general discomfort with modernity and the changes it has brought than simply the result of reading the Bible or studying theology.

Further Reading

Altschuler, Glenn C. *Andrew D. White: Educator, Historian, Diplomat*. Ithaca, New York: Cornell University Press, 1979.

Farrell, John. *The Day Without Yesterday: Lemaitre, Einstein, and the Birth of Modern Cosmology*. New York: Thunder's Mouth Press, 2005.

Griffin, David Ray. *Religion and Scientific Naturalism: Overcoming the Conflicts*. Albany: State University of New York Press, 2000.

Kragh, Helge. *Cosmology and Controversy: The Historical Development of Two Theories of the Universe*. Princeton: Princeton University Press, 1996.

Lambert, Dominique. *The Atom of the Universe: The Life and Work of Georges Lemaitre*. Krakow: Copernicus Center Press, 2015.

Lindberg, David C. and Ronald L. Numbers. "Beyond War and Peace: A Reappraisal of the Encounter between Christianity and Science." *Perspectives on Science and Christian Faith*. 39.3 (1987): 140–149

Conclusion

I started this book by discussing the attitude of students entering my "History and Philosophy of Science" class, so it's fitting that I finish by saying a few words about the attitudes those same students tend to express as they finish my course. As stated earlier, most—like most Americans—enter the class thinking that science and religion are and have always been in conflict. As the semester progresses, I don't often directly address this belief. Instead, I simply provide the students with the historical information about the history of science—including its relationship with religion—as I've sought to do here, while working with them on their critical thinking skills. In the process, most come quickly to see that the historical relationship between religion and science in the West has been positive or neutral far more often than it has been negative. I've had many students state in their evaluations of the course that they were surprised to learn how many scientists have been—and still are—Christians, not to mention how many of these scientists were motivated by their Christian faith to strive for a better understanding of the world.

Few of these students have heard of John W. Draper or Andrew Dickson White before they encounter them in my classroom, but most have been influenced in one manner or another by the ideas these men promoted. Those ideas—centered on the presumed conflict between religion and science—were socially constructed, the product of their times, which included a strong anti-Catholic bias that still exists in many parts of the country. As Philip Jenkins has pointed out in his book, *Anti-Catholicism in America: The Last Acceptable Prejudice*, many Americans not only have a negative view of Catholicism, but they feel perfectly comfortable expressing that view, whereas few would be equally comfortable revealing other prejudices they might hold. In certain regions of the country—such as the South—anti-Catholic views are not only common, they are the norm. I can personally attest to that, as I grew up in Alabama and attended graduate school at the University of Tennessee in Knoxville. As a boy in Alabama I faithfully attended a Southern Baptist Church, and although I don't remember instances in which I was told bad things about Catholicism, I held the general view that it was superstitious and a dying religion. I was shocked as an adult to learn that it's actually the single largest Christian denomination, and I suspect most Southerners would be equally shocked. In Tennessee, as a teaching assistant I graded thousands of papers and exams. In due course of time I taught my own sections of the venerable Western Civilization survey, and it was common for students to write about "Christians" versus "Catholics." As far as many of my students were considered, those were two different categories. Teaching in Wisconsin, I run into that attitude far less often—in part, because there are a lot more Catholics in this area—but it still happens. Far more common, though, are students who flatly say that the Catholic Church has always opposed the development of science. I've even heard such statements from some of my students who are Catholic, who typically feel embarrassed about this sad "fact," but believe it nonetheless.

Anti-Catholicism has an unfortunately long history in the U.S. It reached its peak in the nineteenth century when lurid tales of sexual slavery occurring in convents were printed for salacious readers eager for such stories. This was the era of the first Ku Klux Klan organized marches and night rides to terrify and intimidate Catholics in states from New York to Florida, from South Carolina to California. Things improved from there, but when John F. Kennedy (1917–1963) campaigned for president in 1960, serious concerns were raised about whether he would have the independence necessary to be president, or would he give his allegiance to the Pope over the people of the U.S.? He was the first Catholic elected president, symbolizing an improved status for Catholics in America, but that didn't mean anti-Catholicism was dead. When Pope Paul VI died in 1978, the American evangelist and founder of the university named for him, Bob Jones (1883–1968), wrote an essay titled *The Church of Rome in Perspective,* which begins: "Pope Paul VI, archpriest of Satan, a deceiver and an anti-Christ, has, like Judas, gone to his own place." Most anti-Catholicism isn't this clear, but what the historian Arthur Schlesinger, sr. (1888–1965) once called "the deepest bias in the history of the American people" has never fully disappeared.

Of course Draper and White aren't solely responsible for the idea that religion and science are necessarily antagonistic toward one another. But the fact that their works are still cited, and the anti-Catholicism sprawling across their pages while still often overlooked, is a bit baffling. Furthermore, that modern intellectuals such as Neil deGrasse Tyson and Jerry Coyne continue to reference these writers uncritically is alarming, not only because of Draper and White's rather extreme biases, but also because it's bad practice to rely on historical studies that are more than a century old. A lot has changed and historians have learned a great deal since then.

In spite of direct references made by writers such as Tyson, Dickson and White aren't terribly likely to be the immediate source of the modern cultural belief that religion and science are somehow in conflict, and that belief isn't built simply on anti-Catholicism (though that's still an element). That idea comes from a multitude of often difficult-to-pin-down sources, but is expressed most forcefully by writers such as Christopher Hitchens, Richard Dawkins, or Sam Harris. These writers are as uncritical in their approach to religion as they are forceful in their assertions that it (and by "it" they primarily mean Christianity) has no place in modern society. Because of their confrontational style and uncompromising positions they are often quoted, and have influenced others who sometimes have an even broader reach, such as the television personality Bill Maher (1956–present). Maher has made a career out of attacking religion, which he calls a "neurological disorder," and he has stated that "We are a nation that is unenlightened because of religion. I do believe that. I think religion stops people from thinking. I think it justified crazies." Such statements gain traction in the popular imagination not just because of Maher's prominent position on such high-profile venues as the many specials he has done on Home Box Office, but also because of the anger he engenders among religious conservatives who bring attention to his statements by remarking upon and re-broadcasting them.

The notion that religion—especially Christianity—and science are incompatible is one that is made to seem more plausible by high-profile and vocal opponents of science, such as Ken Hamm (1951–present). An Australian-born American fundamentalist who insists that the Earth is only 6,000 years old, he seems honestly to believe that dinosaurs and humans once co-existed. Hamm promotes his young Earth creationist viewpoints through his organization "Answers in Genesis" and The Creation Museum in Petersburg, Kentucky. This museum shows humans and dinosaurs living together and contains numerous exhibits designed to refute evolution, scientific geology, and other evidence-based approaches to understanding the world. While the museum has proven to be something of a financial disappointment, almost three million people have visited since it opened in 2007. While some no doubt come due to curiosity over such a strongly polemical site that is clearly out of touch with modern science, most visitors undoubtedly come seeking validation for their own anti-scientific ideas. Given Hamm's profile, and that of others like him, it's easy to forget that young Earth creationism is a minority viewpoint in the United States. True, it's still a strong minority—according to the 2014 Pew Research Center poll cited in the introduction, 42% of Americans believe humans were created within the last 10,000 years and haven't changed significantly since that time. Still, the heads of almost all major religious organizations—including leaders of the Catholic Church, the Episcopalian Church, the Presbyterian Church (USA) and many others—have publicly expressed support for the teaching of evolution and opposition to the teaching of creationism in schools. Particularly among Americans under thirty familiarity with creationism is in decline as acceptance of evolutionary theory is on the rise.

Furthermore, it should be remembered that the fundamentalist Christian rejection of scientific views of the formation of the Earth and the development of life emerged relatively recently—only since the late nineteenth and early twentieth century—as the product of historical and social forces, rather than the natural outgrowth of theological views. Religious leaders like George McCready Price and Richard Archer Torrey were alarmed at the fast pace of change that industrialization was bringing to the world and were reacting against modernity. Opposition to scientific views was simply the most visible symbol of that rejection of modernity. Ironically, many of their religious views—such as an insistence that the Bible be accepted as literally, word-for-word true—were quite innovatory and out of step with many centuries of Christian belief and practice.

However, considering how vocal and politically well-connected many modern fundamentalists are in the United States, it makes sense that large numbers of Americans—including many scientists—believe the views they express are those of Christians in general, or a genuine expression of authentic Christianity. This is especially true given the limited knowledge about the history of science that most scientists demonstrate. It's also likely why scientists such as Neil deGrasse Tyson and Jerry Coyne gravitate toward Andrew White's woefully out of date and tremendously biased *History of the Warfare of Science with Theology in Christendom* rather than the more modern and more nuanced work of scholars such as Ronald Numbers (1942–present). White tells scientists such as Tyson and Coyne what they want to hear, affirming what they already think to be true, thus playing to the confirmation bias that is so strong within all of us.

If this promotion of the individual bias that science and religion are inherently in conflict only affected the individual beliefs of these scientists, the problem would be a small one.

Unfortunately, that's not the case. Richard Dawkins has proven to be a tremendously talented communicator, with his 2006 book *The God Delusion* reaching number four on *The New York Times* hardcover non-fiction best seller list. It has sold more than three million copies and presents a rejection of theistic beliefs as a scientific approach to the world, leading many to believe that religious beliefs are inherently opposed to logic. Furthermore, as I've stressed, Sagan and Tyson have reached hundreds of millions more with their two iterations of *Cosmos*, driving home the argument again and again that the history of science is one of repeated conflicts with illogical and poorly conceived religious ideology.

As I stated at the beginning of this book, I have great respect for the work of both Carl Sagan and Neil deGrasse Tyson, and found both versions of *Cosmos* to be very enjoyable and frequently enlightening. However, as a historian I can't help but be alarmed by their promotion of poorly founded historical beliefs, just as I can't help but find works such as Dawkins that attempt to undercut the intellectual rationale for religious faith more than a little misguided and poorly thought through. Granted, those who accept the New Atheist thinking would say the same of me. The physicist Lawrence Krause wrote an article for the *New Yorker* titled "All Scientists Should be Militant Atheists." He uses the case of Kim Davis, the Kentucky county clerk who refused to issue marriage licenses to gay couples because of her religious conviction that such marriages are against what she perceives as God's law, as a jumping off point in order to suggest that religion is dangerous. He then goes on to quote the biologist J.B.S. Haldane who stated in 1934 that his "practice as a scientist is atheistic," meaning that when he sets up an experiment he operates under the assumption "that no god, angel, or devil is going to interfere with its course." Krause goes on then to state that science and religion have no relationship, evidenced by the fact that in a 30-year professional career, he's never heard the word God mentioned in a scientific meeting. For Krause, religion is "enforced ignorance" and it's the duty of the scientist to speak out against religious beliefs.

There are a number of problems with Krause's line of argument, though. For one, his emphasis on those who oppose modernity based on their religious beliefs focuses on their religion as the issue, when in reality their opposition to and fear of a changing modern world is at least as strong a factor in their rejection of scientific thinking. After all, according to a 2017 Pew Research Center poll, 68% of all American Catholics and 63% of mainline (meaning non-fundamentalist) Protestants accept that humans have evolved over time—only slightly lower numbers than the 77% of "nones" who accept the same thing. Not all Christians are as reactionary as Kim Davis (in fact 57% of both Catholics and mainstream Protestants support gay marriage). Furthermore, the fact that scientists don't mention God at professional meetings isn't evidence of general lack of belief on the part of these scientists, or of any conception on their part that science and religion are at odds. Georges Lemaitre didn't mention God in his work, and criticized—politely, and gently—the Pope for suggesting the Big Bang theory had religious implications, but the Belgian priest certainly had no lack of faith. Furthermore, while scientists as a group are less likely to believe in God than the average American, a slim plurality (51%) say they believe in a deity or a higher power, and that no doubt includes many of the people who have sat alongside Krause at professional meetings. They don't mention God because when doing science as scientists such references are neither necessary nor appropriate, since religion deals with issues of faith and morality, not the processes of the natural world. But that doesn't mean they all believe God has nothing to do with those processes. A 1997 survey found that 40% of biologists, mathematicians, physicians, and astronomers believe evolution has occurred through the guiding influence of God. No doubt many of these scientists agree with the Virginia Tech botanist and Darwin scholar, Duncan Porter, who states, in reference to God and evolution: "As an Episcopalian, I don't compartmentalize those things . . . I put them together in an overall view."

Porter isn't alone in this belief, even if most scientist see no reason to discuss the congruence many see between their faith and their work as scientists. The atmospheric physicist Sir Robert Boyd (1922–2004), founder of the "Research Scientists' Christian

Fellowship" and father of the British space program, which he oversaw for three decades, certainly agreed. In addition to his numerous scientific publications, Boyd also frequently wrote on the interrelated nature of science and religion. Perhaps his most eloquent expression of his faith in both science and religion as uncompartmentalized parts of a whole was the poem he published in 1975 that began,

'In the beginning', long before all worlds
    Or flaming stars or whirling galaxies,
Before that first 'big bang', if such it was,
    Or earlier contraction; back and back
Beyond all time or co-related space
    And all that is and all that ever was
And all that yet will be; Source of the whole,
    'In the beginning was the Word' of God
The Word of God; Reason, Design and Form,
    Intelligence, Whose workshop spans the stars
Expressed within the cosmos and alike
    In what seems chaos; He Who works as much
In randomness as order, Who to make
    Man in His image scorns not to create
By patient evolution on a scale
    Of craft divine which dwarfs a million years.

There can be no doubt that Boyd was a man like Kepler, who saw God's handiwork everywhere in the universe he studied.

That was an attitude the Nobel Prize winner, Charles H. Townes (1915–2015), certainly held. In 1966 Townes bemoaned "scientific absolutism in our thinking and attitudes," and rightly pointed out that the history of science is one of repeated over turnings of what were once thought of as certainties. For that reason, his essay on the "Convergence of Science and Religion" stated that faith is essential to scientists in order to even get started in their work, and argued passionately for a convergence between science and religion. Many might disagree with him, but as the preceding chapters have shown, the relationship between science and religion in the West has more often been one of symbiotic convergence than the conflict that all too many take to have been the norm.

It would be simple to continue to multiply the instances in which scientists make statements of faith, or argue for a congruence between their faith and their scientific beliefs, to the point of tediousness. However, it is no more my intent to argue in favor of a theistic approach to science than it is to suggest that such theistic beliefs are accurate based on the statements of scientists who also happen to be people of faith. Instead, what I mean to suggest is that statements such as those by Krause and Dawkins that religious beliefs are somehow rendered invalid or made ridiculous by science simply can't be sustained in the face of so many scientists who believe otherwise. Nor has the work of any scientist shown that a faith-based view of the world is untenable. Krause might argue this in works such as his *A Universe From Nothing*, but as David Albert, a philosopher of science at Columbia University who also happens to hold a Ph.D. in physics has said of Krause's book (which claims that quantum field theories can alone explain the origins of the universe), "Krauss is dead wrong and his religious and philosophical critics are absolutely right." As Albert puts it, Krause's book seems to be driven more by an intense dislike of religion as, in Krause's view, just plain dumb. However, in Albert's words,

> it ought to be mentioned, quite apart from the question of whether anything Krauss says turns out to be true or false, that the whole business of approaching the struggle with religion as if it were a card game, or a horse race, or some kind of battle of wits, just feels all wrong.

The reason why is simple: it's not a binary choice, either to hold religious beliefs or to believe in the methodologies and findings of science. As long as science has existed, many have managed to do both, and many still do.

In the end, I would argue that it's important to understand the historical relationship between science and religion in the West as it actually existed simply because it's important to aim at truth, rather than grasping firmly to biases. However, it's also important because, as with my Catholic students who feel embarrassed by what they believe has been an antagonistic relationship between their faith and science, too many people think Christianity and science are inherently in conflict. That sets up a situation in which those who embrace modernity, who feel that science is an extremely valuable tool for knowledge production, feel called upon to make a choice between their religion or their acceptance of science. Not only does such a feeling lead to unnecessary psychic distress, it also sets up a false dichotomy, as evidenced by the long history of compatibility between science and religion in the West. It's certainly possible that Townes was too overreaching in his essay when he calls for a convergence between science and religion while arguing that their approaches to understanding the world are actually very similar. But the historical record indicates that for many centuries far more scientists have taken this approach, seeing their religious faith and their pursuit of science as symbiotic, rather than the approach of those who believe these two domains of knowledge are incompatible.

Index

About the Author

Scott E. Hendrix received his Ph.D. in 2007 from the University of Tennessee under the very capable and generous guidance of Tom Burman. Scott's areas of specialty are medieval and early modern science and the history of ideas. His first book, *How Albert the Great's Speculum Astronomiae Was Interpreted and Used by Four Centuries of Readers*, received the "Professor D. Simon Evans Prize for Outstanding Contributions to Medieval Studies" in 2010. Since then, Scott has published half a dozen books and dozens of peer-reviewed articles, book chapters, and encyclopedia articles about medieval and early modern science, beliefs about witches, and cultural theory. Scott teaches at Carroll University in Waukesha, WI, where he lives with his wife, Kelly, and their small menagerie of dogs and cats.

Made in the USA
Monee, IL
17 September 2023

42807762R00163